Electric Motors and
their Controls

Electric Motors and their Controls

AN INTRODUCTION

Tak Kenjo

*Professor in the Department of Electrical Engineering and Power Electronics,
University of Industrial Technology, Kanagawa, Japan*

UNIVERSITY PRESS

OXFORD

UNIVERSITY PRESS

Great Clarendon Street, Oxford OX2 6DP
Oxford University Press is a department of the University of Oxford.
It furthers the University's objective of excellence in research, scholarship,
and education by publishing worldwide in

Oxford New York

Athens Auckland Bangkok Bogotá Buenos Aires Calcutta
Cape Town Chennai Dar es Salaam Delhi Florence Hong Kong Istanbul
Karachi Kuala Lumpur Madrid Melbourne Mexico City Mumbai
Nairobi Paris São Paulo Singapore Taipei Tokyo Toronto Warsaw

and associated companies in Berlin Ibadan

Oxford is a registered trade mark of Oxford University Press
in the UK and certain other countries

Published in the United States
by Oxford University Press Inc., New York

© Tak Kenjo, 1991

Reprinted 1999

British Library Cataloguing in Publication Data

ISBN 0 19 856240 3

Library of Congress Cataloguing in Publication Data
Kenjo, Tak
Electric motors and their controls: an introduction/Tak Kenjo.
Translated and updated from Japanese book ABC of motors.
Includes index.
1. Electric motors.   2. Electric motors—Electronic control.
I. Title.
TK2511.K46 1991   621.46—dc20   90-21544

Printed in Great Britain by The Bath Press, Bath

Preface

Some 40 years ago, when I was young, there were three electric motors in my home. One was DC with a U-shaped permanent magnet that drove the propeller on my model motorboat. Another was the small motor in my electric train, which was supplied by a single-phase network via a variable-voltage transformer. This was an AC series motor, but it could also be run on a DC battery. It seemed obvious that these motors should, and did, have two terminals. The third was an induction motor my father used to power the timber saw in his factory. When it was placed on a concrete floor before installation, I noticed that there were three terminals, so I asked my uncle which one was positive, which was negative, and what the third one was for. He may have tried to explain three-phase alternating current, but all I remember now is that I didn't understand a word of what he mumbled.

A couple of years later, a small motor-repair workshop opened nearby to rebuild the burned-out motors from surrounding factories, tea-refining mills, and machine workshops. I used to visit the master and his two apprentices to watch their work. They wound coils on wooden frames to replace old ones, varnished them, and so on. My biggest question was: why didn't these large motors have brushes and commutators like the ones in my models? The rotors were just a massive piece of iron with a closed copper circuit that looked a little like a squirrel cage. Once I asked the younger apprentice how electricity was supplied to the rotor circuit without any contacts, but he was not exactly a Faraday so I didn't get a satisfactory answer.

The only book I owned on motors at the time was really only for hobby use, so even though I read and reread it I couldn't find anything on three-phase induction motors. I searched several bookshops in my town in vain, and it wasn't until after I had started studying electrical engineering in college that I understood the basics of AC motor theory.

I have an interesting experience from the very early stages of my career as an engineer that I would like to share. During lectures on electrical machinery for final-year students, there was a topic on a new type of motor that, although only briefly discussed, I understood very clearly, while most of my peers seemed to have paid no attention. In 1964, after completing my postgraduate work in plasma physics for a masters degree, I went to work for the Teac corporation. On my first day, the president asked me to improve the hysteresis motor driving tape-recorder capstans. This was the very motor whose principle I had fully understood from that brief lecture! I was almost certain that I could create an excellent motor for Teac, and after many hours of hard work my new design was put into mass production for use in the high-class audio equipment being sold to American GIs on their way to the Vietnam War.

As the small motor was not yet regarded as a proper engineering subject in Japan, I decided to write the first engineering paper on my design idea in English. Although it later became a major design used in at least three companies, it was rejected by the Institution of Electrical Engineers, so I rewrote it in my poor third language, German, to be published in a journal in Munich.

After that, I worked to expand my understanding of various types of electrical motors and electronic controls. The theoretical product of this endeavour was consolidated into a dissertation in 1970. However, theory alone did not seem to be enough so I created an experiment bench called the Mechatro Lab that helps explain the interrelationships between various motors, power electronics, and microprocessor controls. This concept and associated subjects were dealt with in one of my books published by the Oxford University Press.

After this, I realized a long-cherished dream, and wrote for the layman. Surprisingly, the Japanese version has been read by a number of generalists in the business sphere as well as many young students.

Allow me to apologize for not writing this English version before the Japanese one, but, as an engineering writer, I have come to recognize an enormous gap between my first and second languages. At the same time, I feel that it is also more effective to write in the mother tongue first and then revise it in English afterwards to produce a better result. In this, John Mullins, who studied Electrical Engineering at Oklahoma State University and who was working on his Japanese at the International Christian University in Tokyo, helped me. I would like to express my warmest gratitude to him.

Kanagawa   T. K.
September 1990

Contents

1
How are electric motors used?

First, in this book, we will see how many electric motors are used in homes, offices, laboratories, factories, hospitals, and so on. We will come to realize that our modern lives function only with the help of a large number of small motors, not just the large ones that pull trains or that are found in factories.

1.1 How many motors are used in American and European homes?

Let us find out how many motors are used in a typical American or European home, as the number of small motors used in a house can be thought of as an index to the quality of life. Before doing a detailed count, one might guess that it would be some 20 or so.

Let us first see what it was like in the America of the 1950s. The famous film comedy *The Seven Year Itch* will give a good indication. In this film, produced in 1955, Marilyn Monroe and Tom Ewell played the leading roles. The man who Tom plays is 38 years old and the section chief of a publishing company. Each room of his flat has an air-conditioner blowing cool air on a hot and humid summer night. The model Marilyn plays comes to the flat after an odd little thing happened, and enjoys the cool air from the expensive equipment, flapping her skirt. The fans in these air-conditioners must have been driven by electric motors just like they are nowadays.

As summer is not so hot in Britain, air-conditioners of this sort are not really needed. Summer in Japan is generally very hot and humid, and many new houses have an air-conditioner. However, in the 1950s, no houses or flats had such luxurious things. Only rich families could afford even an electric fan. As stated in the Preface, there were only three motors in the house where the author spent his childhood.

Table 1.1 is a recent list of motors serving Mr Moteur's household. You may regard it as typical of a suburb of New York or Los Angeles. Mr Moteur's house has central air-conditioning; in summer it generates cool air and in winter hot air to each room. Mr and Mrs Moteur have two sons and one daughter. Mr Moteur's hobbies are listening to music and do-it-yourself. They have a desk-top and a lap-top computer. One of their sons likes playing with radio-controlled cars and aeroplanes. Their only daughter plays piano and violin, but there are no motors in her musical instruments.

Table 1.1   Motors used in Mr Moteur's home

(1) *Living*		(4) *Amusements/hobbies*	
central air-conditioner	3	video recorder	3
refrigerator	2 or 3	record player	3
coffee mill	1	CD player	3
food processor	1	cameras	4
electronic range	1	radio-controlled models	4
egg beater	1	power tools	2
dishwasher	1		
vacuum cleaner	2	(5) *Personal computers*	
hair-driers (2)	2	desk-top computer	
electric razor	1	fan motor	1
electric meter	1	hard disk drive	2
ceiling fans	5	floppy disk drive	2
kitchen ventilator	1	printer	4
toilet ventilator	2	lap-top computer	
gardening machines	3	fan motor	1
		floppy disk drives	4
(2) *Clothes*			
sewing machine	1 or 2	(6) *Study appliances*	
washer/drier	3	photocopying machine	1
		pencil sharpener	1
(3) *Timepieces*		facsimile machine	3
clocks	8		
watches	4		

Although in this example the family have about 80 motors, a typical family in the West normally owns more than 60 small motors. Moreover, if the motors in each of Mr and Mrs Moteur's cars are included, the number of motors exceeds 100. This fact shows that living a modern life without motors is unthinkable.

1.2 Motors in domestic life

The motors listed in Table 1.1 are classified into the following categories: living, clothes, timepieces, amusements/hobbies, personal computers, and study appliances.

It is not the purpose of this chapter to discuss principles of electric motors or their classifications; these will be dealt with in Chapters 2 and 3. However, excepting the ultrasonic-wave motor, it will be appropriate to note that there are three large categories of motors as follows:

1. AC motors: the motors which are driven on the single-phase or three-phase commercial network of 60 or 50 Hz. AC stands for alternating current.

2. Conventional DC motors: the motors with two terminals which are connected to the two terminals of a battery. However, DC (or direct current) motors are often operated on a DC supply converted from an AC net. A structural feature of the DC motor is that it has a copper commutator and carbon or metal brushes.

3. Electronically controlled precision motors: the major members of this category are the brushless DC motor and stepping motors. Brushless DC motors are normally used with speed control, and stepping motors are suited for position control.

Particular types of motors tend to be mainly used in each of the application categories. Most motors falling into the category 'living' are AC motors known as squirrel-cage induction motors driven on a single-phase net. However, AC commutator motors, which have brushes like DC motors, are used in coffee mills, egg scramblers, and vacuum cleaners, because this type of motor produces high speed and a high starting torque for a small size. However, the universal motor, unlike the squirrel-cage induction motor, is not suited to long running times.

Much of the equipment in a typical kitchen, from the refrigerator to the food processor, uses electric motors. Small DC motors are also used in hairdriers and electric razors.

Motors are also used in the clothes category: washing and drying machines and electric sewing machines.

The category 'timepieces' is next to 'living' in terms of the number of motors. To move the arms in most of the latest 'quartz' watches, a tiny permanent magnet is motion-controlled by a crystal oscillator, powered by a small battery. Needless to say, digital-display watches do not use a motor. Watches with a spring and gears instead of an electric motor are extensively used in developing countries, however.

One of the latest trends is the popularization of computers and office automation equipment. For a floppy disk drive, one brushless DC motor and one stepping motor are used. In a hard disk drive, too, a brushless DC motor and a stepping motor are used. However, the so-called voice-coil motor explained in Chapter 6 is used in a high-response hard disk drive in place of the stepping motor. Both brushless DC and stepping motors, unlike ordinary AC and DC motors, require an electronic circuit to control their motion. The brushless motor revolves the memory disk at a constant speed with analog feedback control. The stepping motor is, by means of digital technique, for position control of the magnetic head for writing or reading information on the disk. Even a compact lap-top computer has two brushless and two stepping motors (for the two sets of disk drives) and a brushless fan motor for device ventilation.

The category of amusements and hobbies uses various kinds of motors.

Devices dealing with information, such as video recorders, record players, and CD players have brushless DC motors. Most cameras use plain conventional DC motors. A radio-controlled car is also interesting because it carries three motors and several batteries. The largest battery, which can be used repeatedly by charging, is to power the DC motor driving the wheels. To actuate the speed-control lever, a small motor with a gearhead is used. Another small DC motor is for steering. These small motors are powered by dry cells and controlled via radio signals. A powerful DC motor also turns the propeller of a model plane. Servomotors controlling the elevator and rudder are much smaller than those used in a radio-controlled car. A universal motor is used in do-it-yourself power tools. In a camera, one motor is used to wind the film on and some for other functions. One of Mr Moteur's cameras carries an ultrasonic-wave motor for quick autofocusing.

1.3 Electric motors in automobiles

Most automobiles, except trucks and buses, have and always will have four wheels. However, the number of motors in a car is increasing. Figure 1.1 shows some examples of their applications. When starting the engine, a so-called starter motor is driven. The two windscreen-wipers are operated by a DC motor. The radiator cooling fan is also driven by a motor, and the fuel pump is run by a motor, too. These are very necessary motors, but others are used for power windows, door-locking, sunroof operation, mirror-position control, and so on. Some luxurious cars carry close to 100 motors.

Most motors used in a car are conventional DC motors. These motors are powered by batteries and do not need complicated electronics, and hence they are cheap recently. However, electronically controlled stepping motors have been used in some applications, such as controlling the ratio of fuel to air before injection into a cylinder. A small stepping motor is used in the odometer and also in the clock.

The way the car wheels are controlled in four-wheel steering applications is an interesting example: both front and rear wheels are used in conjunction. At low speeds, the rear wheels turn in the opposite sense to the front wheels to make the rotating radius small. Recently however, when the speed is high, the rear wheels are steered in the same direction but at a smaller angle. If a motor is used for steering the rear wheels, control can be carried out by a powerful stepping motor governed by a microprocessor.

A car is designed as an extension of a human body, and is expected to respond to the engine-speed and steering commands of the driver so that it will not collide with anything. If compared to a human body, the petrol engine or the electric motor in an electric vehicle are analogous to the muscles. An electronic system including one or more microprocessors assesses the driver's

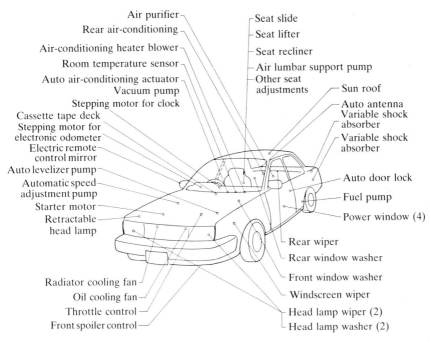

Air purifier
Rear air-conditioning
Air-conditioning heater blower
Room temperature sensor
Auto air-conditioning actuator
Vacuum pump
Stepping motor for clock
Cassette tape deck
Stepping motor for
electronic odometer
Electric remote
control mirror
Auto levelizer pump
Automatic speed
adjustment pump
Starter motor
Retractable
head lamp

Seat slide
Seat lifter
Seat recliner
Air lumbar support pump
Other seat
adjustments
Sun roof
Auto antenna
Variable shock
absorber
Variable shock
absorber
Auto door lock
Fuel pump
Power window (4)

Radiator cooling fan
Oil cooling fan
Throttle control
Front spoiler control

Rear wiper
Rear window washer
Front window washer
Windscreen wiper
Head lamp wiper (2)
Head lamp washer (2)

Fig. 1.1 Small motors in an automobile.

intention and sends control signals to the motors. In this sense, the control technology of car motors will be more and more important in the future.

1.4 Motors in information equipment

Now, let us look at the motors used in applications other than those of household appliances for a while. It is often said that we are in the midst of the info-society. Personal computers, word processors, printers, and lap-top computers are seen everywhere: in offices, banks, government agencies, laboratories, and factories. Facsimile machines are quite common as well. These are the machines that deal with information, and small precision motors are used in all of them. Here are some examples.

1. *Floppy and hard disk drives.* Magnetic disks are used to store large amounts of information from a computer or word processor. There are basically two different kinds of magnetic disks: floppy and hard disks. The base material of a floppy disk is a flexible plate coated with iron-oxide powder as the magnetic recording medium. A hard disk is made of metal so that it can be run at higher speeds. The disks are rotated by a brushless DC motor, and

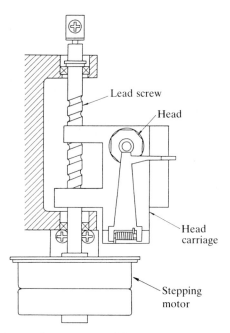

Fig. 1.2 A stepping motor used for magnetic-head positioning.

data is written and read by a magnetic head controlled by a stepping motor or something similar as illustrated in Fig. 1.2.

2. *Printer and graph plotter*. There are various kinds of printer used for typing out documents, program listings, or data. Figure 1.3 shows a serial printer known as a daisy-wheel printer. This is an automatic typewriter that prints one character at a time. A character wheel carrying letters and numerals is directly coupled to a motor mounted in a carriage. There are several motors used in this type of printer. Recently, laser printers have become widely used, because they feature fast, quiet operation and high-quality print with a variety of typefaces other than roman fount style. In a Japanese-language word processor, thousands of characters must be dealt with and the documents are printed out using a laser printer. In a laser printer, a polygonal mirror is rotated at a high speed to scan the reflected light on to a drum as illustrated in Fig. 1.4. The drum has a photoconductive layer on its surface. The latent image of the information to be printed is formed on the drum surface by the laser and then developed by the attracted toner. The developed image is then transferred to normal paper and fixed using heat and pressure. The drum is usually controlled so as to run at a constant speed while in operation.

3. *Facsimile machines*. The device designed to transmit documents or drawings to distant locations, even to overseas countries, via telephone lines is

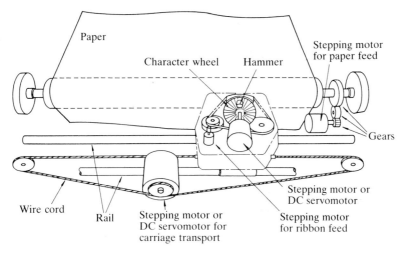

Fig. 1.3 Motors used in a daisy-wheel printer.

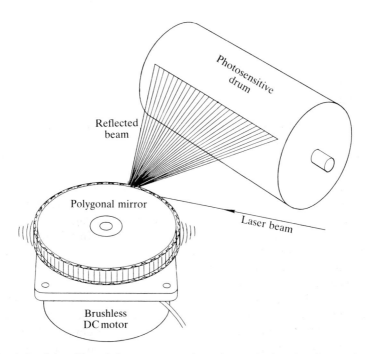

Fig. 1.4 A brushless DC motor turns the polygonal mirror in a laser printer.

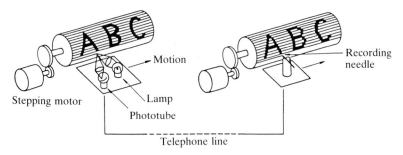

Fig. 1.5   Principle of the facsimile (fax) machine.

called a facsimile, or fax, machine. The basic principle of a facsimile machine is illustrated in Fig. 1.5. The document or drawings wrapped on the sending unit drum is scanned in both the horizontal and rotating directions. The document is divided into graphic elements which are converted into electrical signals by a photoelectric reading head. The signals are then sent over public telephone lines to the receiving unit. The signals received are demodulated and reproduced by a recording head. Motors are used to drive the reading and recording heads. Also another motor is used to drive the drum.

4. *For ventilating electronic equipment and office machines.* It is surprising to learn that an enormous number of motors run propellers to cool electronic and information equipment, such as various types of personal computers, photocopiers, slide projectors, and overhead projectors. The simplest is an AC motor known as a shaded-pole induction motor (see Fig. 1.6). So-called

Fig. 1.6   Shaded-pole type of induction motor (Courtesy Sungshin Co., Ltd.).

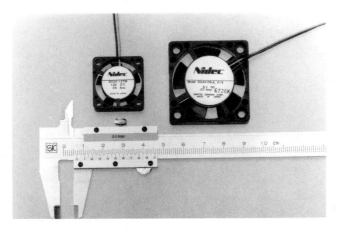

Fig. 1.7 Miniature cooling fan using a brushless DC motor (Courtesy Nidec Corporation).

brushless DC motors are usually used to turn miniature fans as shown in Fig. 1.7.

1.5 Motors on the factory floor and in robots

The reader will already know that motors are essential for the running of manufacturing plants. In iron mills, the motion of heavy rollers is controlled by DC or AC motors in hot or cold strip plants. Lathing machines, milling machines, and index tables use lots of small and medium-size motors as shown in Fig. 1.8.

The first robot to start work in a factory was installed in 1961 at a General Motors plant in the USA to unload hot pieces of metal from a die-casting machine. Many factories have recently introduced lots of automated machines and robots to increase productivity and release workers from arduous or unpleasant work. Control motors drive these machines. By 1989 there were more than 30 000 robots in the world. Japan is now the world leader in robot use and manufacture. When Mrs Thatcher, the former British Prime Minister, visited the Fanuc plant built at the foot of Mount Fuji, robots were manu-facturing more robots at unmanned sites. Robots are absolutely necessary for operation in dangerous environments like nuclear reactors.

How many motors are used in a robot? Some robots can have more than ten (Fig. 1.9). The arms and finger motion must be quick, and hence robot motors must be designed with special considerations. Motor operation is carried out under the control of artificial intelligence.

If even one of the motors in a robot stops or burns out, the production line

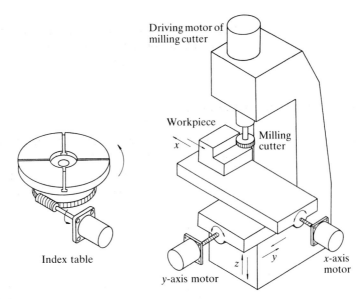

Fig. 1.8   Machine tools such as index tables and milling machines use small and medium-size motors.

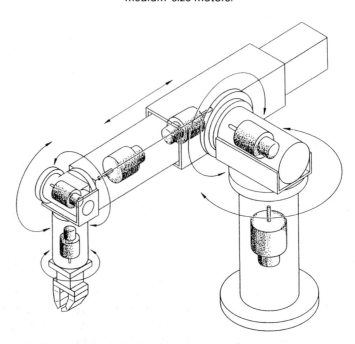

Fig. 1.9   A robot uses many motors.

would stop. At worst, if the drive system of a motor works incorrectly, it could hurt a worker nearby. Consequently, motors and control systems must be absolutely reliable.

1.6 Automatic vending machines

If 'robot' means a tool doing something instead of a human being, ticket machines at railway stations, and vending machines for drinks, tobacco, or magazines are robots. Small motors are used in these devices as working servants. Cash dispensers in banks and supermarkets use many motors as well.

1.7 Transportation

Most railway trains are driven by large DC or AC motors. These motors are normally rotary. However, linear AC motors have begun to be used since linear motors make the train more compact and suited to underground carriages.

Milk floats in British towns are electric vehicles, where a DC motor is installed instead of a petrol engine. An electric vehicle designed for use as a normal passenger car would have to be very heavy because of the large lead battery required to run it several hundred kilometres without recharging or replacement.

1.8 Motors in toys and amusement machines

The reason why there are no toys except for a radio-controlled model car in Table 1.1 is because Mr Moteur's family has no little children who play with toys. However, if the sons and daughter used to like toys, and their toys such as cars and dolls that move by themselves have not been thrown away, and kept somewhere in the house, the number of motors will surely increase. Many recent toys are battery-driven and use a motor.

Amusement machines such as fruit machines or one-armed bandits or Japanese 'pachinko' machines use motors. The most well-known Chinese game, mah-jong, is played with the help of motors. Some ten motors are used in a mah-jong machine for shuffling tiles and lining them up on the table ready to start a game for four people. Surprisingly, a motor is used for throwing the dice to decide the person to start the game. Motors may also be used for playing cards.

1.9 Motors in vision and sound equipment

A lot of families have recently purchased video recorders. These incorporate several motors to move the tape and rotate the cylindrical magnetic head, and to deal with the loading and unloading of the cassette tape.

Machines which produce images use precision motors for various purposes. Figure 1.10 shows a camera stand for special effects. Sixteen stepping motors, which are suited to digital control, are controlled simultaneously by many microprocessors. The camera position, focusing, and positions of several original illustrations with reference to the camera are automatically controlled by software while in continuous filming.

Sumo wrestling is photographed from various angles. A small, light camera hangs from the shrine-shaped roof to follow the quick motion of the fighting

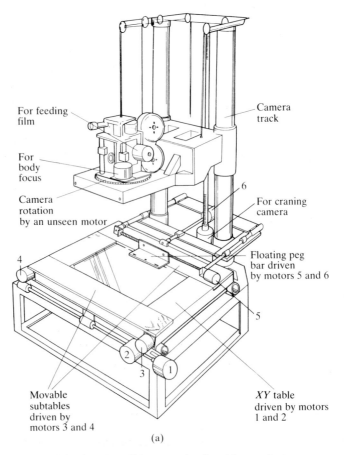

For feeding film

Camera track

For body focus

Camera rotation by an unseen motor

6

For craning camera

4

Floating peg bar driven by motors 5 and 6

5

2

3

1

Movable subtables driven by motors 3 and 4

XY table driven by motors 1 and 2

(a)

Fig. 1.10   Special-effects stand using 16 stepping motors.

wrestlers. Interestingly, the motors in this camera system are controlled from the Tokyo control centre via telephone lines, even when the match is fought in Nagoya. The same technique can be applied to the filming or recording of other sports as well.

Tiny motors are used in compact cameras: for autofocusing, shutter operation, winding the film, and so on. Record players or CD players use several motors too. A camera which features quick autofocusing employs the latest motor design known as an ultrasonic or supersonic motor.

1.10 Motors in medical and healthcare equipment

Various motors are used in medical equipment. Let us look at some examples.

1. *Motors in a dentist's drill.* We used to have decayed teeth whittled off with a drill driven by a long string belt. The belt was powered by an AC motor installed near the dentist's floor. Later, the motor was replaced by an air turbine. The drill is mounted on a turbine turned by strong airflow provided by a compressor. Now, the latest drill is once again driven by an electric motor. However, this motor is different: not a conventional AC motor, but a brushless DC motor. The drill and motor assembly is similar in size to a pen, which is very easy for a dentist to manipulate.

2. *Artificial heart motors.* There are motors for circulating blood during an operation and for replacing a heart which has almost lost its function. Studies to fit an artificial heart or blood pump in an animal body are being carried out. However, it is a rather difficult task, because a small but powerful motor generates heat, which can burn the muscle tissue surrounding it. It makes us realize what an excellent pump our heart is.

3. *Electric wheelchair.* Electric wheelchairs for the physically handicapped are also driven by motors. Two DC motors generate the motive torque to its wheels, supplied by a battery under the seat.

4. *Others.* Motors are always used for driving machines and circulating air in any piece of medical equipment. For instance, the tension control in the neck-pulling machine for the treatment of the marrow system in cosmetic surgery sometimes employs a DC motor. The air compressor used to clear the pipe connecting ear and nose is also driven by a motor.

1.11 Production growth of small motors

Accurate statistics on the production of electric motors are not available. Figure 1.11 shows the growth and decay of the production of several kinds of

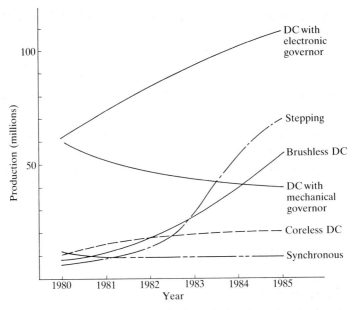

Fig. 1.11   Production growth of small electric motors in Japan.

small motors for control use in Japan. It is said that more than 70 per cent of the world's motors are supplied by Japanese manufacturers. We can see how active the motor industry is, and a collapse of this industry would be inconceivable.

Conclusions

We have seen in this chapter that our life is supported by electrical motors large and small, and that in particular small electrical motors are manufactured in great quantities. It was also seen that small electronically controlled precision motors are essential for running our highly advanced info-society. In the next chapter, we will review the physics of several types of motors.

2
Physical principles of various types of motor

In the preceding chapter, we saw that convenience in daily life is supported to a great extent by electric motors and we could not manage in today's civilization without them. Here, we shall examine some fundamental questions such as: on what principles do these motors operate and why are there so many different types?

2.1 Making a motor with a magnetic needle

Let us take a magnetic needle to start with. From ancient times it has been known that a magnetized iron needle points to the Earth's North and South Poles. With this knowledge, we came to understand that the Earth is something like a large magnet surrounded by a magnetic field. The magnetic needle is caused by this field to point in the direction of the North Pole. That is, when a magnetized needle is allowed to rotate freely as shown in Fig. 2.1, the magnetic north pole of the needle points to the Earth's North Pole, and the needle's south pole to the Earth's South Pole. Here, note that the Earth's magnetic south and north poles are located near the North and South Poles associated with the axis of revolution, respectively, as indicated in Fig. 2.1.

Now, if the Earth's magnetic poles were to move, the needle would rotate to follow the movement. Although the revolving force applied to the needle may be very weak, the needle would be a motor. According to geophysicists, studies of the very weak residual magnetism on rocks at various places on the globe have revealed that the Earth's magnetism must have moved in the course of billions of years, and the magnetic poles have inverted several times to date. Moreover, the north pole was once located near the equator. However, when observed over the span of a hundred years, the Earth's magnetism is relatively stationary and hence the needle is not a motor.

If, in place of the Earth's magnetic field, we provide a strong local magnetic field using a coil of wire and place the magnetic needle in it as shown in Fig. 2.2(a), then we can perform an interesting experiment. When a battery is connected to the coil to produce a current, a magnetic field appears.

Here, before explaining the experiment, the names of several physicists should be mentioned. The first is German physicist Georg S. Ohm (1787–1854) who discovered the famous Ohm's law. When a battery is used to supply

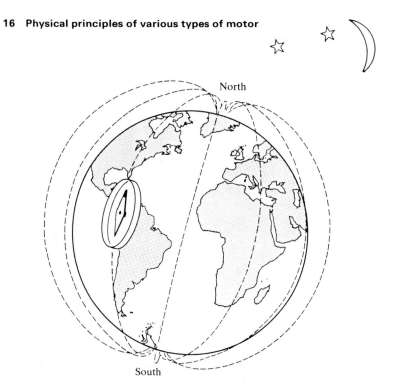

Fig. 2.1   The Earth is a huge magnet with a North Pole and a South Pole.

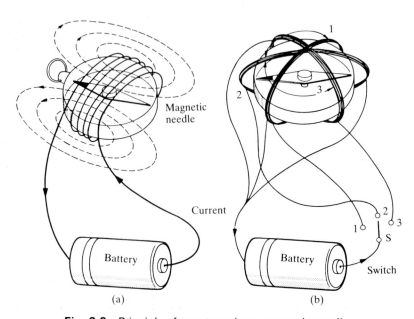

Fig. 2.2   Principle of a motor using a magnetic needle.

an electric current, a flow of electricity, to a circuit, the current I is determined by the formula

$$V = RI, \tag{2.1}$$

where V is the battery voltage, which can be thought of as some force to cause an electric current. One dry cell provides a voltage of 1.5 volts. The unit of voltage, the volt, is named after the Italian physicist Count Alessandro Volta (1745–1827), the discoverer of constant-current electricity. The parameter R is known as the 'resistance' and is specific to a particular circuit. The unit of resistance is the ohm. If the voltage is 1.5 volts and the resistance is 0.5 ohm, the current is $1.5/0.5 = 3$ amperes. The unit of current is named after the French physicist André M. Ampère (1775–1836), who formulated in mathematical terms Danish physicist Hans C. Oersted's discovery that a force acts on a magnetic needle placed near a wire carrying an electric current. It is known that a magnetic field is generated by a current flowing in a wire. Ohm's law does not apply to a circuit to which an alternating current voltage is applied, and instead a more complicated law is used.

The symbols for various units associated with electricity and magnetism are summarized in Table 2.1.

Now let us return to Fig. 2.2. The needle is now exposed to this field and caused to rotate and settle at a right angle to the coil. Next, let us take three coils numbered 1, 2, and 3, respectively, and arrange them at 120° intervals as in Fig. 2.2(b). We place the needle at the centre of the coils, and supply a current to coil 1. Next, we switch on the current to coil 2, and then to coil 3. What happens to the needle? Each time a coil is switched on, the needle revolves through a 120° angle. This is the principle on which a motor having a permanent magnet as its rotor operates. A type of modern stepping motor can be made by further modifying the coil arrangements and using a strong

Table 2.1 Physical quantities, symbols, and units in association with electricity and magnetism

Physical quantity	Symbol	Unit	Unit symbol
Voltage, emf	V, E	volt	V
Current	I	ampere	A
Electric power	P	watt	W
Resistance	R	ohm	Ω
Magnetic flux	Φ	weber	Wb
Magnetic flux density	B	tesla	T
Frequency	f	hertz	Hz
Force	F	newton	N
Torque	T	newton-metre	N m

Note: Electric power is the product of applied voltage and supplied current.

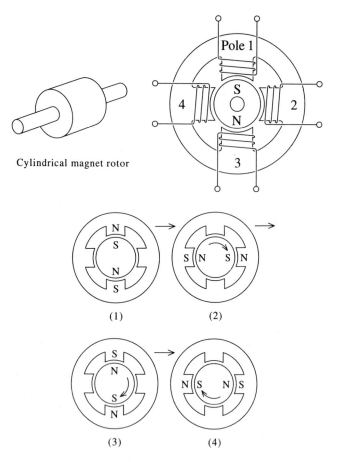

Cylindrical magnet rotor

Fig. 2.3   Principle of a motor using a cylindrical permanent-magnet rotor and four coils.

cylindrical magnet. Figure 2.3 shows such a motor, where four coils are placed at right angles. By switching on the current to the coils in sequence (1, 2, 3, and 4, or in reverse), the magnet rotor will revolve with a step angle of 90°.

2.2 Principles of a stepping motor without permanent magnet

The examples mentioned above in this chapter are motors in which a permanent magnet (a needle or a cylinder) revolves. To see if a motor always needs a permanent magnet, we shall examine an interesting motor originally developed for use in British warships.

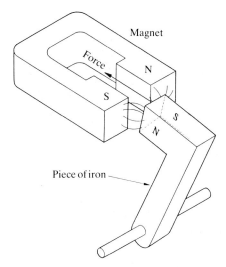

Fig. 2.4 A piece of iron is pulled into the magnet opening.

This motor employs the principle of a magnet attracting a small piece of iron as shown in Fig. 2.4. When the iron is placed near the magnet, magnetic poles are induced on the surface of the piece of iron by the magnet. At the place nearest to the magnet's north pole a south pole appears, and at the nearest place to the magnet's south pole a north pole appears. Unseen magnetic flux bridges the gap between the two different poles. J. C. Maxwell (1831–79) proposed the idea of the magnetic flux and magnetic lines of force to give a mathematical expression to the force created by magnetism. According to his theory, magnetic lines of force act like elastic strings: they tend to shrink themselves. The piece of iron in Fig. 2.4 is being pulled into the opening of the magnet because of the tension along the magnetic lines occurring between the magnet and the piece of iron.

The outline of a motor used in a British warship (Fig. 2.5) was printed in a magazine issued in 1927. Referring to this material, the author designed the test motor shown in Fig. 2.6. We shall note that the most important components in every motor are the stator and rotor. The stator is the stationary component, which is usually fastened to something in a machine, while the rotor is the revolving member, used to move something else. In the motor shown in Fig. 2.6, the outer part is the stator consisting of one iron core and six coils. The inner member, an iron core with four teeth, is the rotor. The stator core in this case has six teeth or salient poles. Any coil and the one opposite it are connected in series, comprising a single electromagnet, normally called a phase. Hence, there are three phases in this motor, as illustrated in Fig. 2.7.

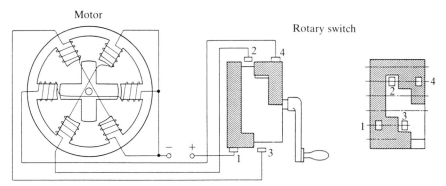

Fig. 2.5   Stepping motor illustrated in *The application of electricity in warships*, issued in Britain in 1927. After World War II, this motor was greatly developed for use in numerically controlled machines in the USA.

Fig. 2.6   An experimental variable-reluctance stepping motor constructed after the model shown in Fig. 2.5.

Current is supplied from a DC battery to the coils via switches. Look at Fig. 2.7. In state (a), the two coils of phase 1 are supplied with a current, which produces a magnetic field at the tooth tips of the first phase. Owing to the magnetic lines of force thus created, these two teeth on the stator and some two of the rotor teeth are brought into alignment. Next, when the current is switched to phase 2, the fields near the first-phase teeth disappear, and new fields appear in the vicinity of the two teeth belonging to phase 2 as shown in

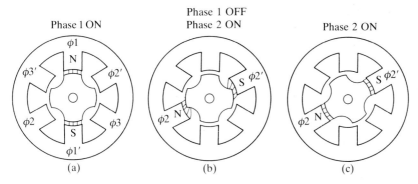

Fig. 2.7 Principle of variable-reluctance stepping motor: (a) magnetic field is excited by current in phase 1 to bring teeth into alignment under phase 1; (b) current is switched to phase 2 and a counterclockwise torque is generated; (c) after a step, teeth are aligned under phase 2.

state (b). By the magnetic field thus generated a counterclockwise rotating force is applied to the other two teeth of the rotor. The rotor will revolve until it eventually comes to the position where these rotor teeth are in alignment with the stator teeth of the second phase as in state (c).

The rotor has travelled 30° in this motion. If the current is switched to phase 3, the rotor will make another 30° rotation. If the switching is operated in the correct sequence like this, the rotor will rotate in a regular manner.

Thus, a permanent magnet is not always necessary in a motor. However, note that the rotor in this case cannot be a smooth cylindrical core and must have protruding parts, called teeth or salient poles, with concave slots between the teeth.

2.3 Principles of a DC motor

Now let us study the principles of a very familiar DC motor. As we saw in Chapter 1, most motors used in cars, toys, and radio-controlled models are DC motors. Therefore the production of DC motors is greater than any other type of motor. There are a number of different variations of DC motors, and these will be discussed in more detail in Chapters 3 and 4. Here, we shall examine a motor having the rather simple structure shown in Fig. 2.8. The stator has a housing and two permanent magnets. The rotor is equipped with copper windings uniformly placed on a cylindrical iron core. The commutator is an important part of the rotor too; current is supplied to the rotor windings via the commutator and two brushes. The brushes are mounted on the stator so as to slide on the commutator surface when the rotor turns.

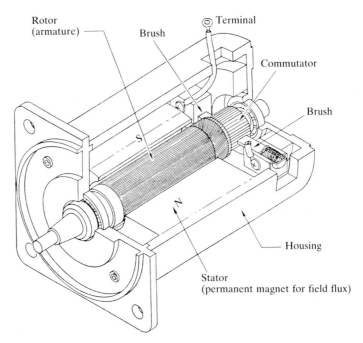

Fig. 2.8   Cut-away view of a DC motor.

It is appropriate to begin with Fleming's left-hand and right-hand rules to explain the principle of a DC motor and highlight its features.

The principle of the force acting on the rotor is given by the well-known BIL law. In Fig. 2.9(a), a conductor, for example a copper or aluminium wire, is placed in a magnetic field. It is known that, if a current flows in the conductor, a force will act upon it. The force F is related to the length L of the conductor, the magnetic flux density B, and the current I as follows:

$$F = BIL. \tag{2.2}$$

The left-hand rule determines the direction of the force relative to the current and field directions. Stretch the thumb, index finger, and middle finger of your left hand as shown in the figure. If the middle finger is the current and the index finger the magnetic field, then the direction of the force is given by the thumb. When the field direction reverses, the force reverses too. However, if the current and field reverse simultaneously, the direction of the force remains unchanged.

In the windings of the motor of Fig. 2.8, the current distribution is as illustrated in Fig. 2.10. If the current flowing in the conductors to the left of the symmetrical axis AB is in the direction away from the reader ($\otimes$), then the current in the conductors to the right flows in the opposite direction of

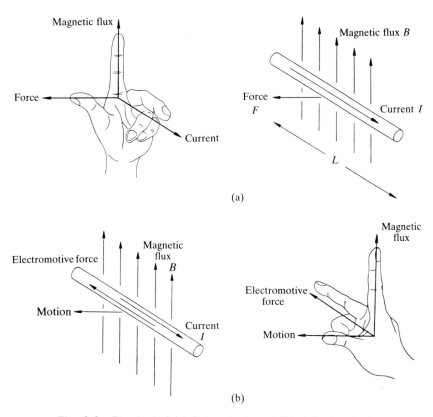

(a)

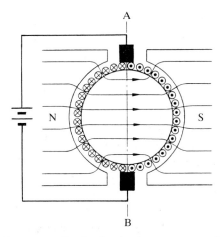

(b)

Fig. 2.9   Fleming's (a) left-hand rule and (b) right-hand rule.

Fig. 2.10   Electric current distribution in the armature coils in a DC motor.

towards the reader (●). The brushes and commutator segments always distribute the direct current from the terminals in the above manner. Of course, if the current is reversed, the distribution will also reverse with respect to the symmetrical axis AB.

In this figure, the conductors in the right half are under the south pole and the conductors in the left half are under the north pole. All the conductors are acted upon by a counterclockwise force according to the left-hand rule, and produce a rotating torque on the rotor shaft.

The characteristics of a DC motor cannot be explained only by the principle of torque production. We need another principle: generation of electromotive force acting on a moving conductor in a magnetic field. Fleming's right-hand rule is associated with this law, and is complementary to the left-hand one.

Let us start with the set-up shown in Fig. 2.11. Two permanent-magnet DC motors, A and B, are coupled by a rubber tube, and a small light bulb is connected to the terminals of motor A. When a DC voltage from a battery is supplied to the terminals of motor B, it will run at a high speed, and the bulb lights. Motor A is being rotated by motor B, and working as a generator to supply an electric current to the lamp.

Here, the term 'generator', also known as 'dynamo', means a device which converts mechanical input power into electric current. In Fig. 2.9(b), a force is working on a conductor in a magnetic field. When the conductor moves as

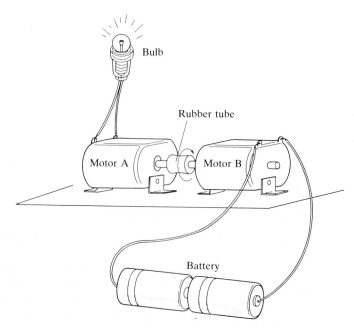

Fig. 2.11 Experiment to show that a motor can function as a generator.

a result of this force, a voltage, which is referred to as an electromotive force, is generated in the conductor. If the speed between the conductor and the magnetic field is v, the electromotive force e is given by

$$e = vBL. \qquad (2.3)$$

The relationship between the directions of field, motion, and electromotive force is given by the index finger, thumb, and middle finger of the right hand as shown in Fig. 2.9(b).

A current powerful enough to light a small bulb cannot be obtained with only one wire unless the magnetic field is very strong or the rotational motion is extremely high. Hence, some way to combine the electromotive force in many coils or conductors is needed. A mechanism to do this will be explained next.

Look at Fig. 2.10 again. Since each conductor passes under the north pole and the south pole repeatedly as the rotor turns, the voltage generated in the conductor alternates direction in accordance with the right-hand rule.

As stated above, there are two brushes and a commutator in the DC motor of Fig. 2.8. They switch current or voltage in association with a sliding mechanism in a motor. When the motor is rotated by an external force, the AC voltage generated in each coil is converted into a DC voltage and combined to a fairly high level by the brush-commutator mechanism.

On the other hand, when a motor operates as a 'motor', the DC current supplied by the battery is converted to an AC current in the rotor by the coordinated mechanism of brushes and commutator. The term 'commutator' is used here because 'to commutate' can mean to convert a DC current to an AC current using a switching mechanism.

2.4 Generator function in a motor

The dynamo used for a bicycle light is one of the most familiar devices employing the right-hand principle. The wheel mounted on the dynamo shaft rolls against the tyre of the bicycle wheel. When the bicycle runs, the dynamo is driven at a high speed, and generates electricity to supply to the light.

Even in a DC motor, however, the right-hand principle must be taken into account. As explained above, a motor produces torque arising from the interaction between the current supplied from the battery and the magnetic field created by a permanent magnet or an electromagnet. However, electricity is also generated in a running motor because the conductors are passing through the magnetic fields, with the voltage being proportional to the rotating speed. Note that this voltage works against the current direction. In other words, the voltage generated in the motor helps to reduce the current. Therefore, as the torque is proportional to current, it decreases as speed increases.

How fast does the motor run when it carries a load? It is the speed at which the torque generated and the load torque are in balance. Here the term 'load' means a dynamic friction or something similar. For example, when a motor drives the propeller of a model plane, wind friction is generated by stirring the air. So a propeller is a load for the motor. When the plane is rising the air resistance is large and the speed is low. In this condition, the battery voltage is much higher than the reactional voltage generated in accordance with the right-hand rule in proportion to the speed. Hence, a large current flows to the motor and this results in high power consumption. However, when the plane descends, the air resistance becomes small and the motor's speed increases, which boosts the generated voltage, and the power consumption is reduced. Thus, when the load is light, the speed is high and the power consumption is low. Conversely, if the load becomes heavy, the motor runs slowly and the power consumption is high.

When a DC motor runs without a load, very little current flows. In this state, the battery voltage and the generated voltage are in balance with each other. If the battery voltage increases, the current increases to accelerate the rotor. However, when the generated voltage matches the supply voltage the speed settles to a constant value and the current becomes negligible again. The right-hand principle has the economical effect of saving wasteful electric power when the load is light.

2.5 DC motor characteristics

If what we have seen above is summarized using some mathematics, the electromechanical characteristics of a DC motor will become more clear. First, according to the fundamental relationship stated by eqn (2.2), the torque produced in a permanent-magnet DC motor is proportional to armature current I_a. If the concept of a 'torque constant' K_T is introduced, this relationship is given by

$$T = K_T I_a. \tag{2.4}$$

As we shall see below, the torque produced by an induction motor, the most widely used AC motor, is a rather complex function of current, speed, and frequency. Why does such a simple relationship hold for a DC motor? One reason is that the magnetic flux needed to generate torque, which is referred to as 'field flux', is provided by a permanent magnet. One can assume that this flux level does not vary with the speed or the current flowing in the coils. To simplify this, let us assume that each conductor in Fig. 2.10 is subjected to the same strength of field, though the polarities are different between the right and left halves with respect to the symmetrical axis AB. In practice, the current

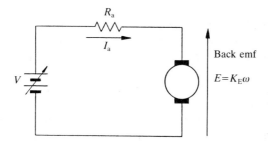

Fig. 2.12 Equivalent circuit of a DC motor.

does affect the flux distribution, and this effect is technically termed 'armature reaction'. However, we will assume that the armature reaction is negligible, because in a well-designed motor it hardly influences performance.

As for the current, since, as will be explained in the next chapter, the same current flows in every coil, the torque applied to each conductor is the same. Thus, the sum of the torques in each conductor is the torque appearing at the shaft, given by eqn (2.4).

Similarly, the voltage E generated inherent in the motor, which is known by the technical term of 'back electromotive force', or back emf for short, is proportional to angular speed ω and given by

$$E = K_E\omega, \tag{2.5}$$

where K_E is the back emf constant.

The torque constant and back emf constant are very important parameters and are always given in motor specifications. Next, let us derive the relationship between torque and speed using these two equations.

Figure 2.12 shows a way of representing a practical DC motor with an electrical circuit called an equivalent circuit. By 'equivalent' we mean that this circuit represents the fundamental properties of an ideal DC motor and sometimes can incorporate other factors associated with the motor. Using this circuit, we can derive the torque–speed relation and the ratio of the loss to the effective power converted to useful mechanical work.

Before deriving the torque versus speed characteristics, the meanings of symbols used in the figure should be explained:

A battery or electric power source for driving a motor (an arrow through this symbol indicates a variable source).

An ideal DC motor with brushes (this may also be regarded as the source of back emf).

A resistor (also represents the resistance in the rotor windings).

The torque can be derived as follows. The battery voltage minus the back emf, $V - E$, is the potential applied to the rotor resistance. On the other hand, the voltage drop across the resistor equals the resistance R_a multiplied by current I_a. Hence we have

$$V - E = R_a I_a. \tag{2.6}$$

From this equation, we obtain

$$I_a = (V - E)/R_a. \tag{2.7}$$

Using eqn (2.4), the torque is

$$T = (K_T/R_a)(V - E). \tag{2.8}$$

By substituting eqn (2.5) into this equation, we get

$$T = (K_T/R_a)(V - K_E \omega). \tag{2.9}$$

If we plot the relationship between torque and speed for different battery voltages (e.g. 1.5, 3, 4.5 V, and so on), the result will be like that shown in Fig. 2.13. Each voltage value is represented by parallel straight lines, and from this graph we find the following.

1. The starting torque, which is the torque at zero speed, is proportional to the battery voltage.

2. The no-load speed, or the speed when the motor carries no load, is also proportional to the battery voltage.

3. The torque decreases with speed, and the slope is $K_T K_E/R_a$, independent of the speed, applied voltage, and torque.

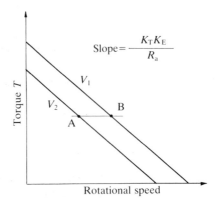

Fig. 2.13 Relationship between torque and speed in a DC motor with a permanent magnet.

The torque constant K_T and back emf constant K_E were derived from different principles associated with the left- and right-hand rules. However, these two constants are closely related to each other. Surprisingly, they are really the same thing in terms of physics. Let us examine this in two ways after making a short comment on the heat loss in a motor.

When a motor, either a DC motor or an AC motor, runs, it gradually gets hot. This is because some of the electric power supplied from a battery or an AC power line changes into heat, and this is called power loss or simply loss. There are, in general, two components of losses: one is loss in the wires in a motor and the other is loss occurring in the core. The former is called copper loss, because the wire is normally copper; the latter is iron loss, as the major material for the core is iron. Note that copper loss occurs in the current path, and iron loss in the magnetic flux path. In a small DC motor, the copper loss is much higher than the iron loss. In the simple equivalent circuit of Fig. 2.12, the copper loss is the loss in resistor R_a and the iron loss is ignored.

If the wire were not made of normal copper but a superconducting material, there would be no resistance in the motor and hence no copper loss. This ideal motor converts all the input power from the battery to mechanical output power available at the motor shaft. When resistance is zero, the battery voltage applied to the motor is always equal to the back emf. Now the input power is EI_a or the back emf times the current. On the other hand, the output power is the work done in one second on the load coupled to the shaft, and this work is the product of the torque T and the rotational speed ω. That is,

$$EI_a = T\omega. \tag{2.10}$$

By substituting E of eqn (2.5) and T of eqn (2.4) into this equation, we get

$$K_E \omega I_a = K_T I_a \omega. \tag{2.11}$$

Dividing both sides by ωI_a, we see that K_E and K_T are equal.

In obtaining this conclusion, we have assumed an ideal condition in which no copper loss or iron loss takes place. This important relationship holds even when appreciable loss occurs in a motor. Let us examine this from a different quantitative approach.

We assume that the magnetic flux has an average value of B. Since, when the radius of the rotor is R, the torque acting on a conductor is R times the force BIL, the torque $RBIL$ works on every conductor and the whole torque T around the rotor axis will be

$$T = ZRBLI = \tfrac{1}{2} ZRBLI_a, \tag{2.12}$$

where Z is the total number of conductors, and I_a is the current supplied from the motor terminal, which is equal to $2I$ because the current is divided into two paths at the brushes.

In this model, the magnetic flux Φ, which is the product of the flux density and the flux area, is equal to

$$\Phi = \pi RLB. \tag{2.13}$$

Therefore, from eqn (2.12), we get

$$T = \tfrac{1}{2} Z(\Phi/\pi) I_a. \tag{2.14}$$

By comparing this equation with eqn (2.4) and noting that $\tfrac{1}{2} I_a = I$, we get for the torque constant

$$K_T = \tfrac{1}{2} Z\Phi/\pi. \tag{2.15}$$

Next, we shall show how the back emf constant K_E can be expressed in terms of other parameters. If the rotor is revolving at a speed of ω radians per second, the speed v of the conductor is

$$v = \omega R. \tag{2.16}$$

Therefore, from eqn (2.3), the back emf E generated in a conductor is

$$E = \omega RBL. \tag{2.17}$$

If the total number of conductors is Z, then the number of conductors in series is $\tfrac{1}{2} Z$ and the total back emf E at the motor terminals is given by

$$E = \tfrac{1}{2} \omega RBLZ. \tag{2.18}$$

By using eqn (2.12) we can express E in terms of the flux Φ as

$$E = \tfrac{1}{2} \omega Z\Phi/\pi. \tag{2.19}$$

Therefore, by comparison with eqn (2.5), we obtain for the back emf constant

$$K_E = \tfrac{1}{2} Z\Phi/\pi. \tag{2.20}$$

This is the same as the torque constant expressed by eqn (2.15).

So, K_T and K_E are the same thing as derived above, but what does this mean?

1. The essential principles of torque production and that of electromotive force generation are closely related and inseparable.

2. The left-hand rule and the right-hand rule are interrelated.

3. If a certain phenomenon is seen from an electromechanical perspective, then it is observed as the generation of force in association with the left-hand rule, and, when seen from another, it is observed as the electro-magnetic induction associated with the right-hand rule.

Unit systems are not always consistent and in such circumstances the torque constant and the back emf constant appear to be different things. The International System of Units (SI) is a self-consistent one. In this system, the torque is expressed in newton-metres and rotational speed is in radians per second. In SI, the unit for the torque constant is newton-metres per ampere and that for the back emf constant is volt-seconds per radian, and these are theoretically the same. However, the English system still used in the USA is a typical inconsistent system. Some metric systems also are not always consistent. For example, in Japanese industry the kilogram-metre is still used as the unit for torque instead of the newton-metre.

In many countries, revolutions per minute (rpm) is frequently used as the unit for rotational speed. This is very out of date. Why should one have to spend an entire minute measuring the speed of a motor in such a high-tech age?

2.6 Induction motors

If one puts a simple cylindrical iron rotor into the experimental stator as shown in Fig. 2.14 and carries out an experiment similar to the one done with the stepping motor, what will happen? It is regrettable that we cannot show this experiment in this book, but the result is that the iron rotor revolves. This is not a stepping motor. Note that the behaviour of the cylindrical rotor is different from that of the salient-toothed rotor. Recall that the latter rotor rotated through 30° and oscillated when settling at a new position. The

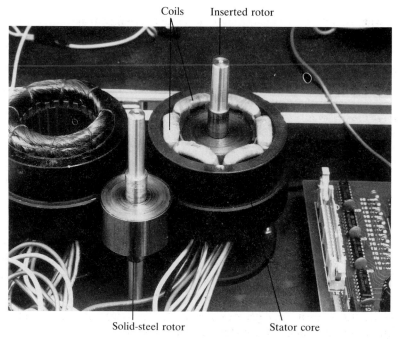

Fig. 2.14 Using a solid-steel rotor for an experiment with an induction motor.

cylindrical iron rotor travels through a smaller angle without the oscillatory behaviour.

When we learned the principles of the non-permanent-magnet stepping motor, we saw that, at switching, the magnetic field produced by the first phase disappears and another field appears in the second phase. This is still true with the cylindrical rotor. Why does such a simple iron rotor work? First, let us change our interpretation of 'switching'. We can think of the magnetic field moving its location from phase 1 to phase 2. The magnetic field travels through the iron rotor: there is a relative motion between the magnetic field and the iron.

From coordinates travelling with the magnetic field, the rotor material moves across the field, and an electromotive force is generated in the rotor core by electromagnetic induction. As a result a current flows in the rotor, which interacts with the magnetic field created by the current in the coil of the second phase. The torque direction is determined by the left-hand rule. Now, for simplicity, let us move directly to the conclusion, skipping detailed calculations: the direction of the torque in this case, too, is the same as before.

If there is a more or less unreasonable assumption here, it is the idea that the magnetic field has moved when it disappears from one place and pops up in another. Yet this is perfectly normal. This is similar to the two side-by-side flashing lights as seen at railway crossings when a train is coming. It looks as if the light is moving from side to side. However, it is true that the movement of the field is not as smooth as that, but much more abrupt, and therefore the torque observed in this experiment is very low compared with the electric power consumed. However, if one improves the stator set-up and rotor construction, the input power utility can be increased. This is the induction motor whose details will be discussed in the next chapter.

Let us review the history of the induction motor before proceeding to new types of motor. Figure 2.15 shows a set-up known as 'Arago's rotation', after the French astronomer D. F. Arago (1786–1853). A copper disc is held so that it can rotate freely, and a U-shaped permanent magnet is placed in such a way that the disc is sandwiched in the magnet opening. Now, if the magnet is moved without touching the disc, the disc will rotate in the same direction.

In 1887 Nikola Tesla (1856–1943) built the first induction motor in America, in which a 'two-phase' alternating current and some fixed electromagnets, instead of a permanent magnet, were used to generate a rotating field.

It was soon discovered that a 'three-phase' AC current is more convenient than two-phase for running an induction motor. A simple model using three-phase current is shown in Figs 2.16–2.18. Three coils are used here, and they are placed in the stator as shown in Fig. 2.16(a) and connected to a commercial power supply as shown in Fig. 2.16(b). Figure 2.17 illustrates the current and magnetic flux distribution when positive voltages are applied to coils A and B, and a negative potential to coil C. Figure 2.18 shows how the current

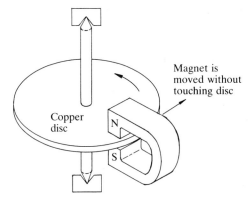

Fig. 2.15 Arago's rotation: the disc turns in the direction in which the magnet is moved.

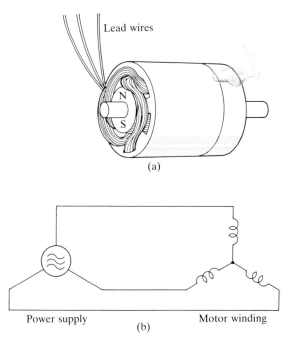

Fig. 2.16 (a) The simplest stator windings of the three-phase induction motor and (b) how to connect the three coils with a power supply.

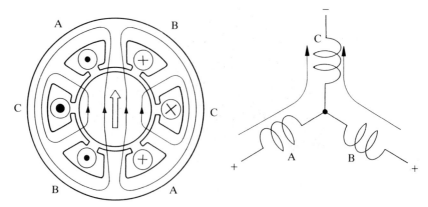

Fig. 2.17 Current and magnetic flux distributions when a positive potential is applied to phases A and B, and a negative to phase C.

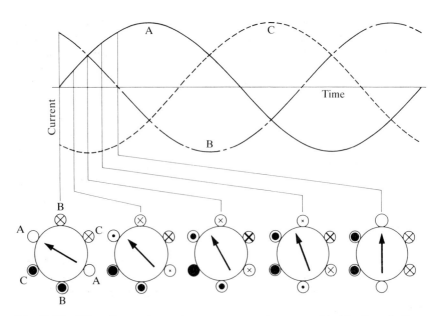

Fig. 2.18 When three-phase sine-wave currents are provided to the windings, the magnetic field revolves at a constant speed with a constant magnitude, which makes the rotor move with little cogging.

distributions vary and how the magnetic field rotates inside the motor when the currents vary in time in three-phase sinusoidal fashion. Thus, when a solid steel rotor is placed in the centre of the stator, it will experience the moving magnetic field and a current is induced in the rotor, which will react with the original flux to yield a torque in the direction of the field rotation.

Stator with   Squirrel-cage
lap windings   rotor

Fig. 2.19   Stator with lap windings for an induction motor and a squirrel-cage rotor.

In this motor, the rotor must revolve at a speed lower than that of the flux so that an electromotive force can be generated by the relative speed between the field and rotor. Thus, to move the magnetic field uniformly, instead of switching the current by means of switches, we use a set of three AC currents varying in a sinusoidal waveform in phases shifted 120° from each other. The stator shown in Fig. 2.19 is a further improvement, and the rotor in this picture is a typical construction of the so-called squirrel-cage induction motor.

Most motors for home appliances are squirrel-cage induction motors, supplied from a single-phase AC source. In these motors, the single-phase current is converted into two- or three-phase currents by some means. More details of the squirrel-cage induction motor will be given in the next chapter.

2.7 The source of torque

Before proceeding further, we should summarize the discussion so far. We have seen a variety of motor principles: two different types of stepping motor, a DC motor with permanent magnets, and an induction motor using no permanent magnets.

However, what is common to all of them is the magnetic field in which something is placed to be rotated. On the other hand, we have seen a variation used in actual machine construction. For example, the rotor in a stepping motor may be a cylindrical permanent magnet or a toothed piece of mild iron; if the rotor is a combination of an iron core and a squirrel-cage conductor, it has the fundamental configuration of the most popular induction motor. We have also learned that there are a variety of coil arrangements.

For further investigation, there are two fundamental ways of thinking. One is an insight in terms of physics and the other is technological. As the latter

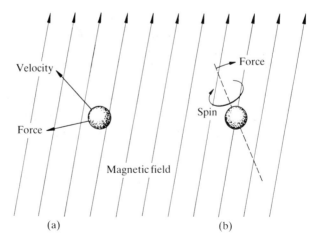

Fig. 2.20 Two kinds of force acting on an electron in a magnetic field: (a) force acting on an electron in linear motion; (b) moment of force acting on spinning electron.

will be dealt with in the next chapter, we shall here discuss the physical approach.

Although we have seen that there are a variety of motors, all modern practical motors are in the last analysis driven by a single basic principle: the force acting on electrons in motion in a magnetic field. However, one may classify the motional behaviour of an electron into two broad categories (see Fig. 2.20):

(1) group motion drifting in one direction in a wire; and

(2) spinning in a fixed space.

The group motion in a wire is an electric current. The interaction between the current and a magnetic field may be regarded as a collection of many interactions between electrons and the field. The forces acting on the electrons travelling in a wire emerge as a force applied to the conductor.

In contrast, the force and torque appearing in a permanent magnet or in mild iron are related to spinning electrons. If it is assumed that the electron's negative charge is distributed over its surface, a force will act on each part of the electron's surface, and the collective force emerges as a force tending to orient the revolving axis in the direction of the magnetic field, as illustrated in Fig. 2.20(b). According to quantum physics, there are only two sorts of spin: positive and negative. One may interpret this as each electron spinning in one direction at a fixed speed or in the opposite direction at the same speed. (Actually, in quantum physics, it is not correct to say that an electron is a particle spinning or revolving around its own axis, but for the sake of an explanation that we can picture we assume that an electron is spinning.)

In most materials, the number of electrons with positive spin and negative spin are the same, and hence there exists no force due to spin. However, in some materials, like iron or cobalt, there is an imbalance in positive and negative spins. In mild pure iron and mild carbon-steel, the spinning axes are in random directions when a magnetic field is not applied, but, when exposed to an external magnetic field, the axes are oriented in the same direction as the field: this phenomenon is called 'magnetization'. The stepping motor using a toothed piece of mild iron utilizes this sort of magnetization to produce torque.

When a permanent-magnetic material is magnetized by a strong external field, the spins are oriented in the magnetization direction. This orientation remains even after the external field is removed, and is little influenced by an external field that is weaker than the level when previously magnetized. As stated, a torque acts on the spinning axis because of an external magnetic field. A permanent magnet, which contains many electrons spinning in the same sense produces the torque in stepping and DC motors.

When viewed microscopically, the torque sources are drifting electrons and electrons spinning in a magnetic field. However, from a macroscopic viewpoint, there seem to be a variety of mechanisms for torque production applicable to motor construction. Both microscopic and macroscopic considerations are very helpful in understanding why a variety of motors can be utilized in our daily lives.

2.8 Laboratory-made fluid motor using thermomagnetic effect

In most practical motors, as we have seen above, the magnetization of mild carbon-steel is controlled by current flowing in coils. However, one can construct a motor without windings by using a phenomenon in which magnetization strength is sharply changed by temperature in a substance, as shown in Fig. 2.21. In this motor, heat energy can be directly converted into kinetic energy, producing a torque or driving force.

Figures 2.21(b) and (c) help to explain this principle. A disc of such material is placed in an air gap between the north and south poles of a magnet. In cross section, the left half of the disc in the gap is kept at a higher temperature than the right side by a heat source. In the portion of the disc at the higher temperature the magnetization is weak, while in the lower-temperature region it is stronger. The flux produced by the magnet has a tendency to bias towards the right because of the difference in magnetization strength. Thus the lines of magnetic intensity are curved. Because of this curvature, a force acts on the disc causing it to travel to the left. When the lower-temperature portion of the plate comes into the higher-temperature region, it is heated and its magnetization is lowered. Hence, as long as the plate or disc moves at a moderate speed, the magnetic-state distribution is maintained in the air gap, and a continuous torque is produced.

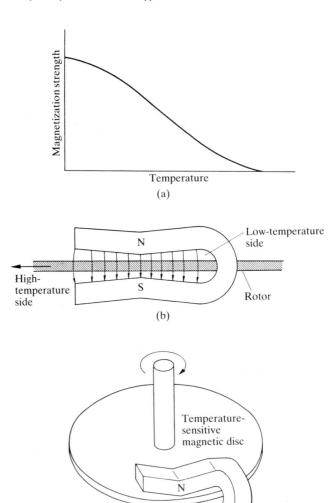

Fig. 2.21 A material whose magnetization (or permeability) varies sharply with temperature can be a rotor: thermal energy can be converted to mechanical work.

The possibility of using magnetic liquid material instead of a solid rotor has emerged for building a new actuator on the principle of thermomechanical energy conversion. The principle of a test motor built in a Japanese Government laboratory is illustrated in Fig. 2.22. The magnetic liquid is water or oil in which fine particles of magnetic material are dispersed to make a colloidal solution. If the characteristics of the material show the tendency illustrated in

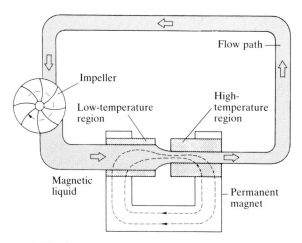

Fig. 2.22   Basic construction of a magnetic-fluid motor.

Fig. 2.21, the liquid flows from the low-temperature region to the high-temperature region.

2.9 Shape-memory alloy motor

There is an interesting family of alloys known as shape-memory alloys. The original shape of this kind of metal is recovered by heating, even after complete deformation at a low temperature. Nitinol and betalloy are well-known shape-memory alloys. Nitinol is an alloy of nickel and titanium; betalloy is made of copper, zinc, and aluminium. Actuators employing the shape-memory principle have been built in several laboratories.

A primitive example of this actuator is shown in Fig. 2.23. A shape-memory alloy wire is used and it is extended by an ordinary spring when it is cold. When the switch is turned on an electric current flows in the alloy and heats it. This heat causes the alloy to contract, recovering its original shape. When the switch is turned off, the alloy will cool and again be expanded by the spring. By repeatedly turning the switch off and on, the pulley will rotate through a certain degree in the clockwise and counterclockwise directions. Because the heat cycle is not very rapid, motional speed is slow with this motor.

2.10 Electrostatic micromotors no wider than a human hair

There is another type of force acting on an electron. This is the force on an electron placed in an electric field, as illustrated in Fig. 2.24. A piece of celluloid

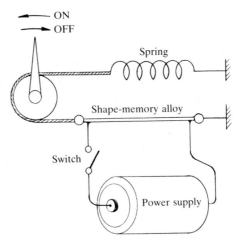

Fig. 2.23 Principle of a shape-memory actuator.

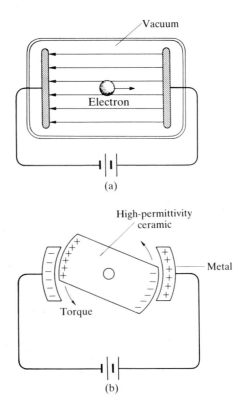

Fig. 2.24 (a) Coulomb force acting on an electron due to an electric field; (b) a torque can be created between the induced electric charge on the surface of a high-permittivity material and electrodes.

rubbed with cloth possesses electricity and will attract pieces of paper or hair. Photocopying machines employ this principle. By the exposure to light, letters or drawings are duplicated on paper, carbon powder or toner is attracted by them, and they are fixed to the page by heat.

However, if one attempts to apply this principle of force generation and build a motor of normal size, the available torque will be extremely low. On the other hand, however, tiny electromagnetic motors are very inefficient. The smallest motor in practical use is the one running a wrist-watch. This uses a very strong permanent magnet about 1 mm in diameter, and the load is only two or three thin arms. If one attempts to build a much smaller motor, most of the electric power will be consumed as heat; the output torque is very low.

It has long been known that electrostatic motors can be advantageous over the normal electromagnetic motor if its diameter is smaller than 1 mm, because the electric charge stored on the surface of a particle is large compared with the particle volume if the particle is small. A team of engineers at the University of California at Berkeley first claimed to have built the world's smallest motor using materials and techniques similar to those employed in the manufacture of semiconductor chips. The construction shown in Fig. 2.25 is similar to the stepping motor of Fig. 2.5; both rotor and stator have teeth but in different numbers. Thus, this latest motor is an electrostatic micro

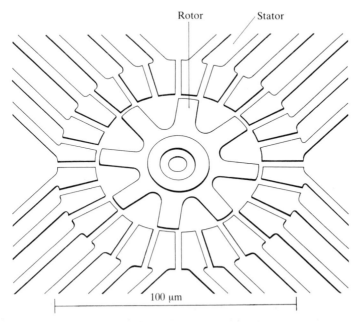

Fig. 2.25   Electrostatic stepping motor built in a laboratory.

stepping motor. The rotor, made of a high-permittivity material, is polarized and electric charges are induced on the tooth tips when a high potential is applied to the stator teeth. A force is generated between the induced charges and the energized stator teeth.

The rotor is 60 μm in diameter, with a thickness of one-thirtieth of that. It will be a long time before this motor will be manufactured at reasonable cost and reliability, and found suitable applications.

2.11 Ceramic motors with minute movement

The inverse piezoelectric effect is an electrostatic effect and much more useful for creating a motor or electric actuator for specific uses. When pressure is applied to a pellet of crystalline material, an electric potential (voltage) occurs. This is known as the piezoelectric effect. Conversely, when a voltage is applied to certain kinds of ceramics, a stress (a form of force) is induced and the ceramic stretches in the direction of the applied field as shown in Fig. 2.26. This is the inverse piezoelectric effect. The ceramic known as PTZ, whose main component is $Pb(Zr-Ti)O_3$, is an attractive material of this type. This ceramic expands and contracts in near proportion to the applied voltage.

Although the strain occurring in these ceramic materials is very minute, an appreciable movement is seen by piling up a number of ceramic layers sandwiching electrodes. Even in this arrangement the movement is small, but the force is fairly large (e.g. higher than 100 N). In Fig. 2.27, a type of printer

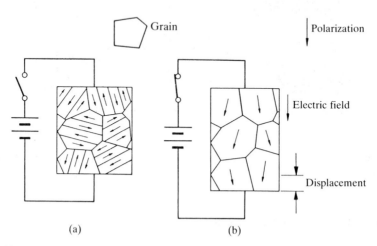

Fig. 2.26 Inverse piezoelectric effect: (a) inside a crystalline grain there are many domains having different polarization directions; (b) when an electric field is applied, the polarization directions align with the field and the crystal expands.

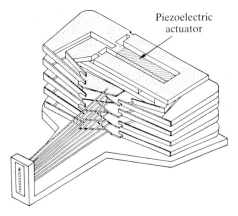

Fig. **2.27** Principle of printer hammer employing the inverse piezoelectric actuator.

Fig. **2.28** Toshiiku Sashida, inventor of the ultrasonic-wave motor.

hammer is shown, where the minute movement is boosted by a lever-and-hinge mechanism.

2.12 Ultrasonic-wave motors

The 'ultrasonic-wave motor' is an attractive new motor invented by Toshiiku Sashida (1939–), a Japanese, from a fundamentally different concept. It also uses the inverse piezoelectric effect, and a kind of dynamic wave travelling in a substance. Figure 2.29 shows the main structure of the new motor and a photograph of the inside.

2.12.1 Standing waves and travelling waves

First, we need to have some knowledge of wave motion, starting with the vibrations occurring in a string of a musical instrument like a guitar, arching in shape as shown in Fig. 2.30. Every part of the string moves up and down

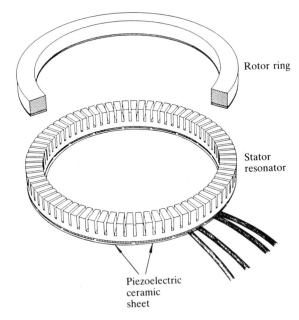

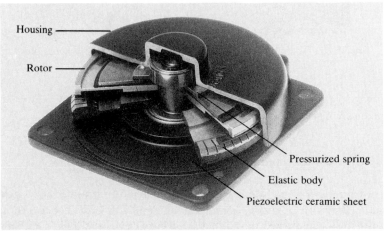

Fig. 2.29   Rotary type of ultrasonic-wave motor.

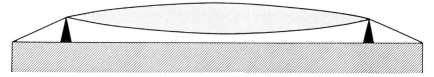

Fig. 2.30 Standing wave occurring on a string instrument like a guitar.

in phase at the same frequency, but the amplitude varies from one place to another. This type of wave is called a 'standing wave'.

There is another kind of wave motion known as the 'travelling wave'. If one throws a stone into a calm pond, waves spread from the place that the stone enters. These waves are typical travelling waves, which we often see. Although standing waves and travelling waves seem to be different phenomena, they are closely related.

Imagine a bathtub filled with water. If one pushes the water in the centre of the tub with a washbowl and removes it at once, a standing wave occurs on the water surface. However, one can interpret this wave as two travelling waves superimposed as follows. As in a pond, two travelling waves are created: one wave travels to the left, the other to the right. They hit against the tub-wall and are reflected to travel in the opposite directions. Then, again, they hit the

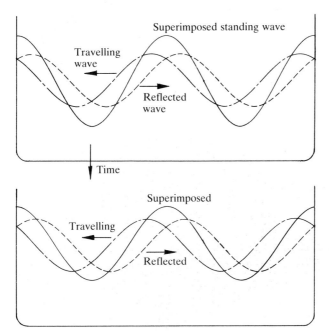

Fig. 2.31 Standing wave occurring on water surface: the standing wave is the overlapping of a travelling wave in one direction and its reflection.

wall and are reflected back, and so on. Thus, in the tub, two travelling waves overlap. This overlapping, or superimposing, creates a standing wave, as illustrated in Fig. 2.31.

In the guitar string, travelling waves can also propagate in both directions and are reflected at the ends where the string touches protruded bars. Since the waves come and go between the two ends, the superimposition creates a standing wave.

2.12.2 Waves in a metal bar

Likewise, when a part of a metal bar is vibrated, a wave will travel in both directions along it. This wave is known as a flexural wave and its characteristics are closely related to the size of the bar and its elasticity. In an ultrasonic-wave motor, the wave travelling in one direction is eliminated, and the other wave is utilized. This wave motion is shown in Fig. 2.32. If one observes the motion of a point on the surface, then, interestingly, the movement is not simply up-and-down, but shows an elliptic locus.

Similar elliptic motion was found in a study by Lord Rayleigh (1842–1919), a British physicist, on a wave travelling in an elastic material, but had not attracted earnest attention until Sashida was inspired to utilize it to make an actuator. This wave is similar to an ocean wave rolling in towards the shore. The movement of a piece of floating wood tossed about by the wave is also elliptic. The wave-head moves towards the shore, and the wave-bottom moves

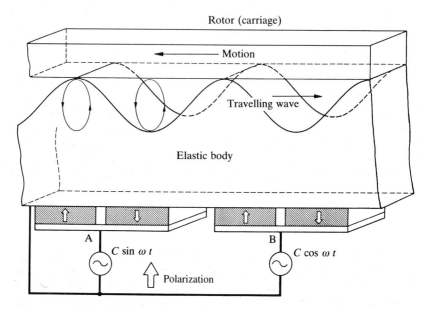

Fig. 2.32 Principle of the ultrasonic-wave motor.

in the opposite direction. That is, the direction of the wave propagation and the movement of the wave-head are the same. This is why the floating pieces of wood are carried to the shore.

However, the metal bar is different: the direction of the wave propagation and the motion of the wave-head are opposite. A water wave seen at the beach and the elastic wave on a metal bar are similar in some ways, but different in others. Since it is not easy to explain this difference and not very necessary, we shall proceed to the mechanism of the ultrasonic-wave motor.

2.12.3 Generating a travelling wave along a metal bar using piezoelectric ceramics

As stated above, we want to generate a travelling wave, not a standing wave. Since a straight bar with a finite length reflects the wave at both ends, a standing wave occurs unless some means is employed to prevent it. If the bar is made into a ring, even if it is of finite length, the same effect as an infinite bar holds. Yet the wave can propagate in both directions. To allow the wave to propagate in only one direction, at least two sources of vibration are needed. Although Fig. 2.32 represents a linear motor model, it can be thought of as a part of a ring used in a rotary motor.

Here, an even number of ceramic sheets are placed at certain intervals, and they are divided into two groups A and B alternately. An electrode is fixed to each ceramic sheet. A sine-wave voltage $C \sin \omega t$ is applied to the electrodes of group A, and a cosine-wave voltage $C \cos \omega t$ to those of group B to generate vibrations. Here, C is the amplitude and ω the angular frequency equal to 6.28 times the frequency f. In this set-up, the surface wave travels to the right. When the voltage applied to group A is $-C \sin \omega t$, the wave direction will be reversed. This ring is used as a stator.

Now, another piece of metal, which is free to rotate, is placed against the stator ring with a high pressure, as shown in Fig. 2.32. When the travelling wave is generated in the stator ring, the rotor ring is in contact with the stator metal at the wave-heads. Since, as explained above, each point of the stator ring is in elliptic motion and the wave-heads are moving in one direction with an amplitude of a few micrometres, the rotor ring receives a thrust in the same direction and rotates (to the left in Fig. 2.32) by the mechanical friction caused by the pressure.

2.12.4 The reason for ultrasonic waves

Why are vibrations in the ultrasonic frequency ranges used for this motor? Ultrasonic waves are defined as acoustic or dynamic waves that cannot be heard by the human ear; quantitatively, this means oscillations higher than 20 kHz. In an ultrasonic-wave motor, mechanical waves are used. As seen above, movement caused by the thrust in one cycle is only a few micrometres.

If it is 2 μm and this motion occurs 20 000 times a second, then the total movement is 4 cm. Thus, vibrations in the ultrasonic ranges are needed in order to attain a usable speed as a motor. To control the speed, one may adjust the amplitude C by adjusting the voltage applied to the ceramics. It is not realistic to vary the frequency over a wide range, because we must select the particular frequency that is suited to size and material properties, because the wave length of the oscillation, which is closely related to the frequency, is not independent of the ring size.

2.13 Monorail ultrasonic-wave motors

Two types of linear ultrasonic-wave motors were also built in Sashida's laboratory. One type is a monorail, as illustrated in Fig. 2.33(a). The two ends of a long metal bar are connected to each other to form an endless stator rail, and on this rail a carriage is placed under pressure by a spring. The principle of the monorail motor is an extension of the ring motor: travelling waves are excited by placing a group of piezoelectric ceramic vibrators on part of the rail and carefully tuning the oscillation frequency. A drawback of this set-up

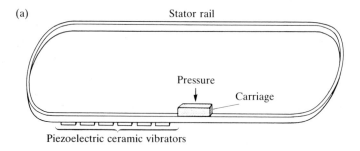

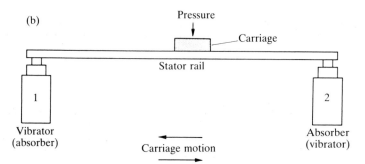

Fig. 2.33   Linear ultrasonic-wave motors.

is that the machine efficiency is poor, because vibration energy is dissipated at every part of the rail.

The other type is literally a linear motor as shown in Fig. 2.33(b); a 'Langevin' type of oscillator is used at each end. This type of vibrator, which also uses piezoceramics, can be an oscillation absorber. When vibrator 1 is used as the oscillator, the waves travel towards end 2, where the vibrator 2 acts as an absorber. In this mode, the carriage experiences a force towards the left. If there is no absorber at this end, the waves are reflected so as to travel towards the left: the result is a standing wave along the rail. When vibrator 1 is operated as the absorber and vibrator 2 as the oscillator, the carriage will run in the opposite direction. In this configuration, the carriage can run as fast as 1 ms^{-1}.

2.14 Coils, inductors, and capacitors

Before proceeding to the next chapter, we should study two more elements that are often used in the drive/control circuits of motors. They are the inductor and the capacitor. The coils of the electromagnets in Fig. 2.2 are typical inductors. A motor winding is also an inductor. When an electric current flows in an inductor, a magnetic field is produced, which possesses a form of energy: magnetic field energy. We can say that an inductor is something that stores magnetic energy.

In contrast, a capacitor is something that stores electric field energy. If two sheets of aluminium foil are wrapped sandwiching a thin insulator film like paraffin paper, they form a capacitor. When a voltage is applied between the two aluminium sheets, electrostatic energy is stored in the narrow space occupied by the insulation film. Besides storing energy, a capacitor has useful functions such as electrical-noise elimination and generation of time-lag electrical signals.

In an AC circuit, inductors and capacitors display useful roles. Table 2.2 summarizes the symbols, units, and basic functions of these devices.

Table 2.2 Symbols, units, and functions of inductors and capacitors

	Inductor	Capacitor
Symbol	⟋ⴰⴰⴰⵑ	╢╟ ╢╟
Unit	henry (usually millihenry)	farad (usually microfarad)
Unit symbol	H, mH	F, µF
Functions	Storing magnetic energy	Storing electrostatic energy
	Smoothing current waveform	Eliminating electrical noise
		Generating time-lag signals

Conclusions

We have seen the physical properties of various types of motor. Most of the current motors are driven by the interaction of a magnetic field and electric currents flowing in windings. In the next chapter, we shall analyse some engineering problems of these practical motors using a different approach.

3
Engineering principles in motor design

In the previous chapter, the physics of various kinds of electrical motor was surveyed, and it was emphasized that most motors run by the torque generated from the interaction between electric currents and a magnetic field. In this chapter, we will deal with these motors from an engineering viewpoint.

Engineering is thought of as the application of physics and chemistry for the benefit of human lives. However, this idea is too simple to deal with sophisticated modern motors often driven by microprocessors and power electronics. Here, we will look at two aspects of these motors. The first is the type of winding and how to supply current to these windings to drive the rotor, and the second is the relationship between rotor construction and motor behaviour.

3.1 Stator, rotor, and windings

Some motors are driven by DC power, some by AC power, and some via switches. However, what is common to all of them is that they are all equipped with windings carrying an alternating current. It is also characteristic of the electric motor to have a stator and rotor. Thus the three major components of an electric motor are the stator, the rotor, and the windings.

3.1.1 Various types of windings

Let us start with the windings in the stator. There is a large variety of winding types. However, they can be classified into two basic categories: distributed windings and concentrated windings. A model of distributed windings is shown in Fig. 3.1. This is an experiment bench designed for the study of the principles of motors, and it can be clearly seen how the coils are installed in the slots in the stator core. Figure 3.2 shows an example of concentrated windings. Each tooth in the core is wound with a single coil.

In order to display both the merits and demerits of both types of windings, the author designed test stators, as shown in Fig. 3.3. Stator A has two sets of distributed windings with 24 slots and the same number of stators. One coil covers several teeth, and each coil overlaps with another. Stator B has six coils in the concentrated configuration.

The stator cores are made with laminations of thin steel plates. The metal

Fig. 3.1   Distributed windings in an experiment bench manufactured by Feedback Instruments Ltd.

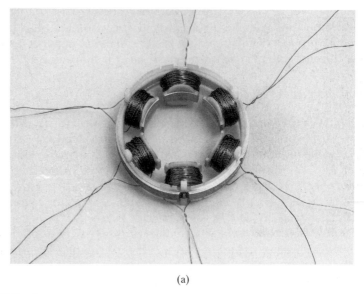

(a)

Fig. 3.2   Examples of concentrated windings: (a) for normal motors; (b, top right) for an outer-rotor motor.

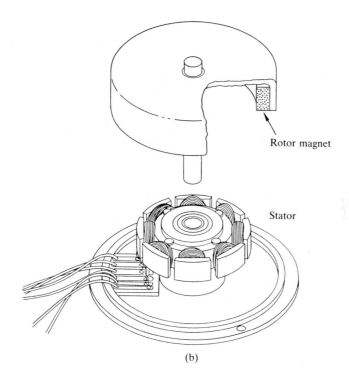

(b)

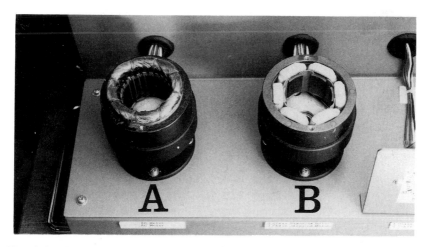

Fig. 3.3 Two stators A and B designed for experiments with the rotors shown in Fig. 3.11.

used is normally silicon-steel or iron containing silicon so as to increase the magnetic flux capacitance. 'Permeability' is the technical term applied to this capacity. Stator cores usually have teeth and slots.

When engineers talk about windings, they often refer to the number of poles. Here 'pole' means magnetic pole. In the examples given in Chapter 2, the number of poles was always two. The magnetic field covering the earth has two poles (one north and one south): this is referred to as a two-pole field. In the stepping motor of Fig. 2.7, the stator generates a south pole and a north pole in each state. Figure 3.4 shows the four-pole windings employed in stator A. When electric currents are supplied to this winding assembly, a magnetic field will appear, as shown in Fig. 3.5. There are two north and two south poles and they can rotate when the currents vary with time, as will be shown later.

Similarly, Fig. 3.6 illustrates the eight-pole windings in stator A, and Fig.3.7

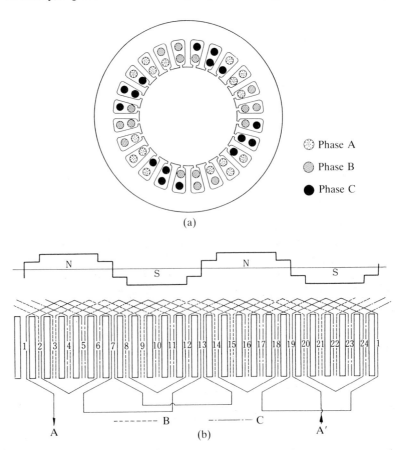

Phase A
Phase B
Phase C

(a)

(b)

Fig. 3.4 The four-pole windings used in stator A in Fig. 3.3: (a) shows how three-phase windings (conductors) are distributed in 24 slots; (b) is the connection diagram for one phase.

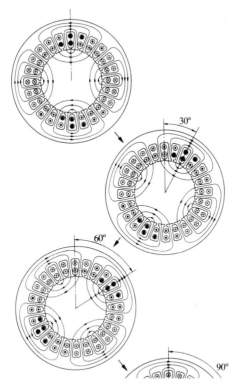

Fig. 3.5 Current and magnetic field distributions in the four-pole scheme; the magnetic field revolves by varying the current in each phase.

shows how four north poles and the same number of south poles appear in this configuration. So, each coil in the four-pole winding covers five teeth and two coils shifted by one tooth pitch from each other consist of one pole, while each coil in the eight-pole winding covers three teeth and only one coil is used for each pole.

The coil arrangement in stator B can produce either a two-pole or a four-pole field, by switching the coil connection, as shown in Fig. 3.8.

3.2 AC motors: classification by rotor structure

When a rotor is placed in the opening of a stator, supported by bearings, and the appropriate alternating currents are supplied to the windings, the rotor will turn like an AC motor. See the structure of such a motor in Fig. 3.9.

Some motors are designed to be powered by a three-phase power supply, and some by a single-phase power supply. Figure 3.10(a) shows how three-phase alternating currents vary with time, and Fig. 3.10(b) shows the two connecting methods.

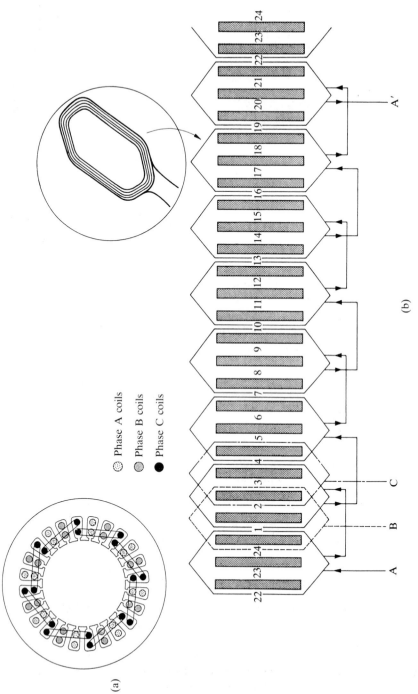

Fig. 3.6 The eight-pole windings used in stator B in Fig. 3.3: (a) shows how three-phase windings (conductors) are distributed in 24 slots; (b) is the connection diagram for one phase.

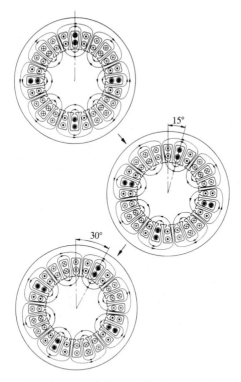

Fig. 3.7 Current and magnetic field distributions in the eight-pole scheme; the magnetic field revolves by varying the current in each phase.

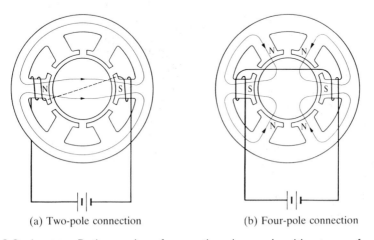

(a) Two-pole connection   (b) Four-pole connection

Fig. 3.8 In stator B, the number of magnetic poles can be either two or four by changing the connections for two opposing coils.

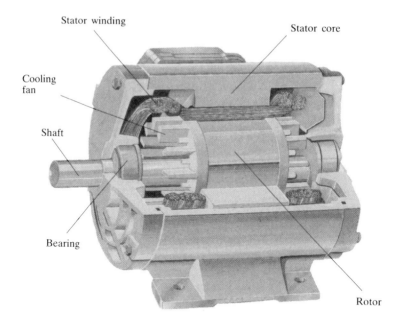

Stator winding

Stator core

Cooling fan

Shaft

Bearing

Rotor

Fig. 3.9 A typical rotor.

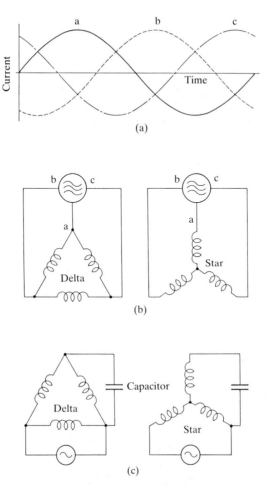

Fig. 3.10 (a) Sinusoidally varying three-phase currents; (b) connecting the windings to a three-phase supply; (c) single-phase operation of an AC motor with three-phase windings using a capacitor.

One is the so-called delta (Δ) connection and the other the star, or Y, connection. Figure 3.10(c) shows a method of operating a motor with three-phase windings using a single-phase power supply. The capacitor and the windings work together here to convert a single-phase current to three-phase current. Since the three-phase current generated in this way is not perfect, it is only used for very low power applications. A similar method is widely used in single-phase motors with two-phase windings as will be seen in Fig. 4.3 (p. 101) in the next chapter.

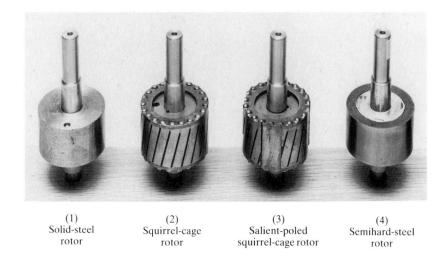

(1)	(2)	(3)	(4)
Solid-steel rotor	Squirrel-cage rotor	Salient-poled squirrel-cage rotor	Semihard-steel rotor

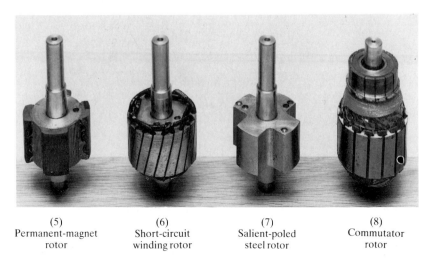

(5)	(6)	(7)	(8)
Permanent-magnet rotor	Short-circuit winding rotor	Salient-poled steel rotor	Commutator rotor

Fig. 3.11   Eight different types of rotors designed to be operated in stators A and B shown in Fig. 3.3.

When three-phase currents are supplied to the windings, a magnetic field is generated and revolves as shown in Figs 3.5 or 3.7. If a rotor is exposed to this field, it is caused to rotate in the direction of the magnetic field.

The relationship between the generated torque and speed, and also some other characteristics, depend on the rotor structure. For this reason, AC motors are named after the rotor's construction method. Eight typical rotor

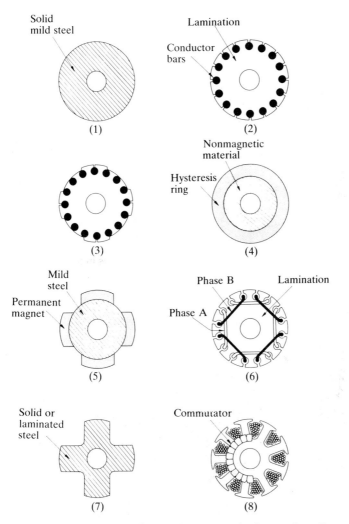

Fig. 3.12   Cross-sections of the various rotors in the previous figure.

designs used in stators A and B of Fig. 3.3 are shown in Figs 3.11 and 3.12.
Let us examine each rotor structure.

3.2.1 Simple solid iron rotor

The simplest rotor is a plain block of iron seen at the extreme left of Fig. 3.11.
The AC motor using this type of rotor is referred to as a solid-steel induction
motor or an 'eddy-current motor'. The relationship between torque and speed
is shown in Fig. 3.13. The torque is high at the start, but decreases with speed.

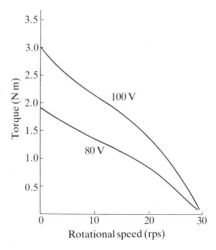

Fig. 3.13   Typical torque-versus-speed characteristics in an eddy-current motor.

Fig. 3.14   Outer-rotor eddy-current motor (Courtesy Papst-Motoren GmbH & KG).

Somewhat complex reduction of characteristics is seen due to a secondary effect of the teeth in the stator core.

The unique motor shown in Fig. 3.14 is an outer-rotor type eddy-current motor: the cup-shaped rotor turns outside the stator core and windings. This motor was used to turn the reels in an open-reel tape-recorder, since it features high torque in the lower speed ranges and low torque ripple.

3.2.2 Squirrel-cage induction motor

The most widely used AC motor has an interesting rotor construction: a squirrel- or hamster-cage-like structure of a conducting material is built in an iron core, as shown in Fig. 3.15. Note that the core is made of laminated silicon-steel. The silicon-steel plate is insulated by a very thin film to prevent undesirable electric current in the core. The conductor material is normally aluminium or copper. Big squirrel-cage motors are used in factories and small ones in home appliances.

The rotational speed of a squirrel-cage induction motor is almost fixed depending upon the number of poles in the windings and the AC frequency. In normal motors, which have four-pole windings, the rotor turns at a speed a little lower than 25 rps (revolutions per second) in the UK and Europe where the AC frequency is 50 Hz. Also, if the load is increased, the speed will decrease. If the same motor is driven in a 60 Hz region, it will rotate at a speed of just under 30 rps.

The test stator A has both four-pole and eight-pole windings. If the same rotor is driven using eight-pole windings, the speed will decrease to about a half of the four-pole speed.

A typical example of the torque-versus-speed characteristics is shown in

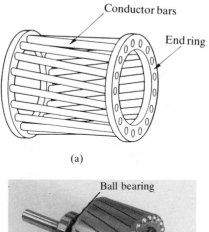

(a)

(b)

Fig. 3.15   (a) Squirrel-cage conductor; (b) squirrel-cage rotor.

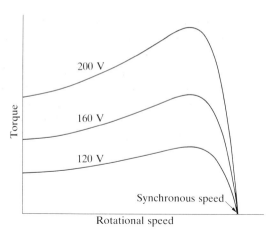

Fig. 3.16   Torque-versus-speed characteristics of a squirrel-cage induction motor.

Fig. 3.16. Note that, if we compare the characteristics of a squirrel-cage motor with those of the DC motor shown in Fig. 2.13 (p. 62), several big differences are seen.

1. The maximum speed is the synchronous speed N_0 that is related to the frequency f and the number of poles p as follows:

$$N_0 = 2f/p \quad \text{(rps)}.$$

This is independent of the applied voltage.

2. Torque increases with speed, reaches a maximum, and rapidly falls away to zero at the synchronous speed.

3. The torque is proportional to the square of the applied voltage at an arbitrary speed.

When operated in the speed ranges around 90–95 per cent of the synchronous speed, an induction motor with these characteristics produces minimum heat losses with respect to the electric power consumed. Hence, the power motors used in factories and small motors for home appliances are designed to operate in these ranges.

When an induction motor is operated with a proper control of both voltage and frequency, useful characteristics are obtained for the control of speed and torque. Such matters will be discussed in Chapters 4 and 5.

3.2.3 Reluctance synchronous motor

Another interesting version of the squirrel-cage type motor is the reluctance synchronous motor. Here, 'synchronous' implies that the motor turns at the synchronous speed in the normal state, independent of the load carried, as long as the load is lower than a certain level. The rotor of the reluctance synchronous motor can be simply a squirrel cage with sections milled out.

The rotor portions which are not milled out are referred to as salient poles. The torque-versus-speed characteristics are as shown in Fig. 3.17. They are similar to those seen in a normal squirrel-cage induction motor, but the unique difference is that, when the speed gets near the synchronous speed, it is rapidly pulled towards the synchronous speed. In this operating state, the magnetic flux distribution will be as shown in Fig. 3.18.

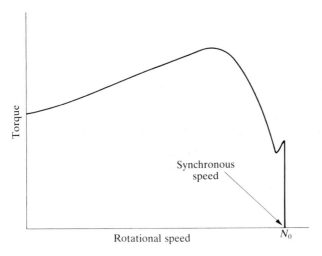

Fig. 3.17 Torque-versus-speed characteristics of a reluctance synchronous motor.

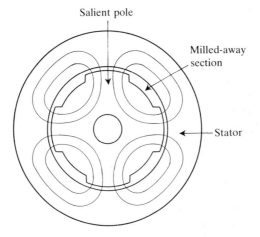

Fig. 3.18 Flux distribution in a salient-pole rotor.

The flux distribution and the rotor move at the same speed, and the flux passes across the smallest gap. 'Reluctance' means unwillingness to do something, and in electromagnetic terms it means magnetic resistance. Magnetic flux is reluctant to pass through the large air gap at the cut-away portions: in other words, a large air gap has a high reluctance. However, magnetic flux shows a tendency to pass through a narrow air gap. This is the principle upon which a salient-poled rotor turns at the synchronous speed. Note that if the rotor has no squirrel-cage conductor, it cannot start and accelerate.

Reluctance motors were once used to drive the turntables of record players. Some hard disks in computer memory peripherals are operated by reluctance motors.

3.2.4 Hysteresis synchronous motor

Another important synchronous motor is the hysteresis motor. If a cylindrical, weak permanent-magnet material without pre-magnetization is used, this is a hysteresis motor.

Hysteresis is a phenomenon in which magnetic flux density and magnetic field intensity are determined by the magnetic history of the material. Figure 3.19 shows a typical magnetic hysteresis; the hysteresis motor usefully employs these characteristics for effective torque production. Before the motor starts it is not necessary for the rotor material to be pre-magnetized. When AC currents are supplied to the windings, magnetization occurs as a result of the field intensity created by the current. If the material is magnetically soft and displays very weak permanent-magnet characteristics, the available torque will be very low. On the other hand, if the material is magnetically very hard, it can hardly be magnetized by the stator current, and consequently the torque is too low. Hence, the rotor material must have magnetic characteristics intermediate between soft and hard.

One advantage of the hysteresis motor over the reluctance motor is that it can be operated at any number of poles. For example, if the stator has windings for two-, four-, and eight-pole configurations, the motor runs at the synchronous speeds 60 (50), 30 (25), and 15 (12.5) rps provided that the frequency is 60 (50) Hz. Such motors used to be used in open-reel tape-recorders. On the other hand, the reluctance motor is basically a single-speed machine. Tiny hysteresis motors are used in gyroscopes, which find important use in aircraft for indicating a reference position in attitude control.

The history of the development of the hysteresis motor is very interesting. The book *Theory of Electricity* by G. H. Livens, published by Cambridge University Press in 1918, carries a brief mathematical expression from which the torque due to magnetic hysteresis is derived. It is said to be C. P. Steinmetz who studied widely the influence of hysteresis on motors, generators, and transformers, and pointed out clearly the possibility of a hysteresis motor.

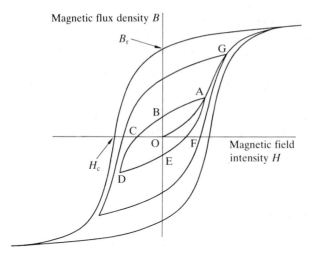

Fig. 3.19 Magnetic hysteresis curve seen in the relation between magnetic field intensity H and magnetic flux density B. The flux is due to the field intensity, which is proportional to the winding current. In air, plastics, and nonmagnetic materials, B is proportional to H. In a permanent-magnet material, however, the B/H relation displays a hysteresis. Initially, the B/H state is in the unmagnetized state denoted by O. When H is applied by the current and increased, B will increase along the so-called maiden curve and eventually reach point A. As the field intensity subsequently gradually decreases from this point, the B/H characteristics do not follow the original curve; instead, the flux decreases along the curve AB. When the field intensity is zero, a certain level of flux density remains. When the intensity is increased with the reverse polarity it will follow BCD. At point D, the field intensity has the same but opposite value as point A. When H is again decreased to zero and increased with positive polarity as before, the B/H state follows another curve DEFA. The closed loop plotted as ABCDEFA is called a hysteresis loop. If the field intensity is increased even more, the B/H state will again follow the maiden curve. A larger loop will be followed if the field intensity is reversed after reaching the point G. When magnetic saturation occurs, the hysteresis loop will not become any larger than a certain size. The value of B at the point where the maximum hysteresis loop intersects the ordinate is referred to as magnetic remanence B_r, while the value of H at which the curve intersects the abscissa is referred to as the coercive force H_c.

B. R. Teare developed a useful theory of hysteresis motors, produced one, and presented the details in a paper in 1940.

How this motor came to be mass produced in Japan is interesting when seen in connection with the popularization of tape recorders and the firm foundation of the small electric motor industry. The AC biasing magnetic recording method, the most important principle upon which a tape recorder operates, was invented in both the USA and Japan during World War II.

Fig. 3.20   Outer-rotor hysteresis motor manufactured by Papst-Motoren GmbH and KG.

When, after the war, American tape recorders came to Japan, engineers were astonished at the rugged hysteresis motor driving the capstan at a constant speed. Because of the war-time isolation, Japanese engineers had little knowledge of the new motor. There were some young engineers who showed enthusiasm in producing usable hysteresis motors to drive Japanese-made tape recorders. Mr K. Tani, who later founded the Teac Corporation and S. Tamura who was with Sony Corporation, are among them.

In 1955, T. Yazaki at Tohoku Metal Industries was successful in creating a cast magnet having characteristics suitable for the hysteresis motor. Motors using this material began to be produced in factories in various places in Japan. The present author was engaged in the study of the optimal magnetic properties of this cast magnet relative to the magnet dimensions and stator parameters. Figure 3.20 shows a German-made unique outer-rotor hysteresis motor.

The reason for the hysteresis motor's rapid decline is that, around 1970, the limitations in the wow/flutter level of this motor when used to drive the rotating head of a video recorder became very apparent. Smaller, more efficient brushless DC motors began to be manufactured to replace hysteresis motors.

3.2.5 Permanent-magnet motor

If a strong permanent magnet is used for the rotor, it is called a permanent-magnet synchronous motor and has a large torque for the rotor size. However, the rotor cannot start if a voltage frequency of 50 or 60 Hz is applied, because it generates a field that rotates too quickly. One starting method is to use an electronic device called an inverter, which provides AC currents of variable frequency. The motor is started at a low frequency and accelerated by increasing the frequency.

Another drawback with this motor is an irregular fluctuation in speed that often occurs during normal operation. For this reason, the permanent-magnet motor is more widely used as a sophisticated brushless DC motor, dealt with in Section 3.7, rather than a simple AC motor .

3.2.6 Wound-rotor induction motor

Some large AC motors have a rotor with windings. The simple rotor (6) in Fig. 3.11 can rotate, but it is really too simple. A scaled-down model of a large practical construction is shown in Fig. 3.21(a). Slip rings are mounted on the rotor to form a circuit with external elements. This rotor has a Y-connected three-phase configuration, and can be used either as an induction motor or a synchronous motor.

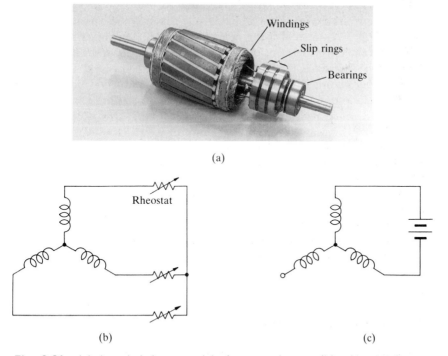

(a)

(b)   (c)

Fig. 3.21 (a) A scaled-down model of a wound rotor; (b) when used as an induction motor, external resistors (rheostats) are connected through slip rings; (c) when used as a synchronous motor, a DC current is supplied through slip rings.

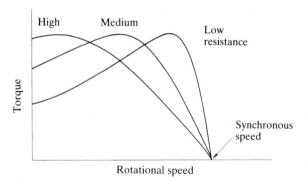

Fig. 3.22   Torque-versus-speed characteristics can be controlled by adjustment of the resistance in the external resistor. Compare with the characteristics shown in Fig. 5.16 (p. 137), where both frequency and voltage are variable.

When used as an induction motor, the circuit shown in Fig. 3.21(b) is formed: a resistor is connected in series with each phase through a brush. The resistor resistance can control the torque-versus-speed characteristics as shown in Fig. 3.22; the higher the external resistance, the higher the starting torque and the lower the consumed current at starting. On the other hand, the lower the resistance, the higher the torque in the high-speed ranges. This induction motor, formerly known as the wound-rotor induction motor, is now of historic interest only, because electronic operation of a simple squirrel-cage induction motor provides better characteristics than control of the external resistance with a wound-rotor induction motor.

When the circuit configuration of Fig. 3.21(c) is employed, the motor becomes a synchronous motor which is widely used. In this arrangement, the rotor is an electromagnet instead of a permanent magnet.

3.2.7 Salient-poled steel rotor

This rotor is seldom used as an AC motor, but is used as a stepping motor. The reader is referred to Section 3.4.

3.2.8 Commutator rotor

The AC motor that uses a rotor with windings and a commutator is called an AC commutator motor or AC series motor. The characteristics of this motor are quite different from other types of AC motors, and in fact it is rather similar to a DC motor. As AC series motors can also be driven on a DC power supply, they are also known as 'universal motors'.

3.3 Revolving magnetic field generation in the stator

The stator of an AC motor generates a magnetic field that revolves in it, and the rotor rotates under the influence of this field. Since the fundamental principle was explained in Chapter 2, here we will see how a magnetic field can be rotated, using a manually operated three-phase inverter.

The two-pole configuration of the stator B windings provides a simple three-phase stator which is very suitable for this purpose. We connect the six coils as indicated in Fig. 3.23. Three motor terminals are connected to the three output terminals of the manual switching circuit as shown in Fig. 3.24. If the current directions in the coils are changed with the three switches in the sequence shown, the magnetic field revolves in 60° steps.

Figure 3.25 shows the switching sequence and voltages applied to the terminals in tabular form. The time variation of the voltage differences between the two phases are also given in the table and graph. In this operation, voltages applied to the windings vary in a step waveform. However, when the switching frequency is moderately high, the electric currents do not vary in a step wave as shown in Fig. 3.26: the waveforms are continuous, though not as smooth as a sinusoidal wave. Consequently, the magnetic field, which is caused by the currents, revolves rather smoothly, not in 60° degree steps.

If three-phase currents transmitted from a power station are supplied to the motor, the magnetic field revolves at a constant speed. This is the conventional way of driving an AC motor.

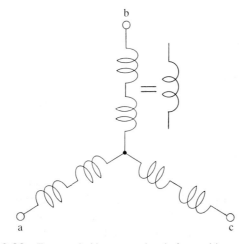

Fig. 3.23 Two-pole Y-connection is formed in stator B.

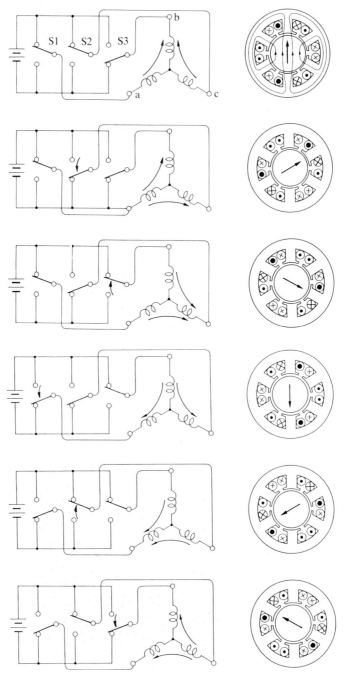

Fig. 3.24 Magnetic field is generated by connecting to a DC power supply via three manual switches as shown, and rotated by operating the switches in a proper sequence.

Switching sequence		1	2	3	4	5	6	7	8	9	10	11
Manual switches \ up / down	S1	/	/	/	\	\	\	/	/	/	\	\
	S2	/	\	\	\	/	/	/	\	\	\	/
	S3	\	\	/	/	/	\	\	\	/	/	/
Terminal voltages	a	0	0	0	E	E	E	0	0	0	E	E
	b	0	E	E	E	0	0	0	E	E	E	0
	c	E	E	0	0	0	E	E	E	0	0	0
Potential differences	a−b	0	−E	−E	0	E	E	0	−E	−E	0	E
	b−c	−E	0	E	E	0	−E	−E	0	E	E	0
	c−a	E	E	0	−E	−E	0	E	E	0	−E	−E

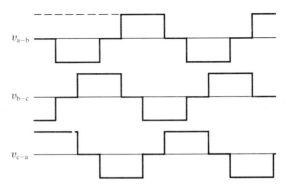

Fig. 3.25   Relationship between switching sequence, terminal voltages, and line-to-line voltages in the inverter operation.

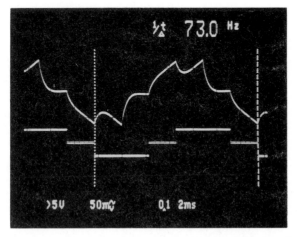

Fig. 3.26   Current waveforms at a phase when the hysteresis rotor is operated at 73 Hz.

3.4 What is a stepping motor?

The seventh rotor in Fig. 3.11 has a very strange shape. If it is mounted in stator B, it is a stepping motor as already seen in the previous chapter. However, it is a special type of stepping motor called a 'variable reluctance motor', or just a VR motor. Another type is the permanent-magnet stepping motor seen in Fig. 2.3 in Chapter 2.

Since the first numerically controlled milling machine using three VR motors appeared in the USA in 1957, this type of stepping motor has been manufactured in large quantities until DC servomotors began to be used in their place. Figure 3.27 shows a multi-stack power stepping motor used in numerically controlled machines, where one stack corresponds to one phase. This motor is a five-phase motor. The principle of a three-stack motor is shown in Fig. 3.28.

As seen in Chapter 2, the stepping motor features a simple positioning control in units of the stepping angle. C. L. Walker, a British civil engineer, invented a way of decreasing the stepping angle by cutting smaller teeth in the pole as shown in Fig. 3.29. Figure 3.30 is a modern design of this type. Walker probably did not benefit from his design, as his invention had been made too long before it was put into practice.

A tiny permanent-magnet stepping motor is used to drive the arms in a watch (see Fig. 3.31). Use of a strong permanent magnet features a high power-conversion ratio and allows the machine to be small. However, a

Fig. 3.27   Stator and rotor of a five-stack VR motor (Courtesy Minebea Co., Ltd.).

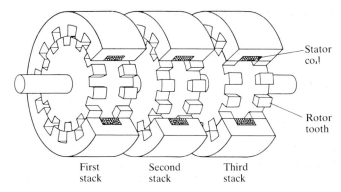

Fig. 3.28 A construction of a multi-stack VR motor.

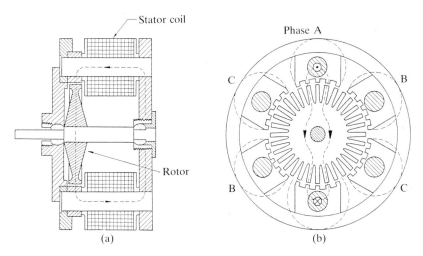

(a)   (b)

Fig. 3.29 A three-phase stepping motor invented by C. L. Walker.

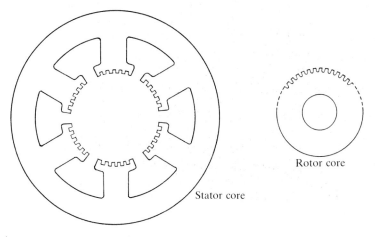

Fig. 3.30 Cross-section of a modern VR stepping motor with a small step angle.

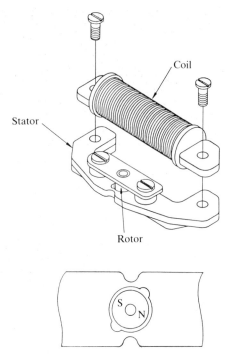

Fig. 3.31 Stepping motor driving the arms in a 'quartz' watch.

drawback with this motor is the large stepping angle, for example 90° or 180°. The hybrid stepping motor, a combination of the permanent-magnet and VR types, illustrated in Fig. 3.32, features both high efficiency and small stepping angle.

Figure 3.33 shows another type of hybrid stepping motor with a thin disc magnet as its rotor. This one was invented by French engineer C. Oudet.

There are also both linear and planar hybrid stepping motors. We shall explain the principle of the hybrid mechanism by reference to a simple linear motor of the type shown in Fig. 3.34, known as the Sawyer linear motor. The motor, which is called a 'slider', consists of a permanent magnet and two electromagnets A and B. The magnetic flux associated with the permanent magnet forms a closed path through the electromagnet core A, the air gap, stator core, air gap again, and electromagnet B.

In the absence of currents in the coil, the flux does flow through both core teeth as shown in electromagnet B in state (a) or (c). When the coil is excited, however, the flux is concentrated at one tooth as shown in electromagnet A in (a). This brings the flux density in this tooth to a maximum, while that in the other tooth becomes negligible.

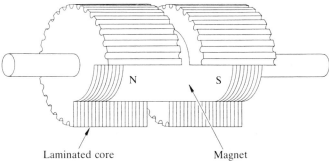

Laminated core   Magnet

Fig. 3.32 Hybrid stepping motor.

Now, in state (a), tooth 1 of electromagnet A is aligned with one of the stator teeth. When current is switched to coil B, in the direction shown in (b), the slider will be driven right a quarter pitch to bring tooth 4 into alignment with the adjacent stator tooth. Electromagnet B is then de-energized and A is excited in the opposite polarity to before. This produces a force bringing tooth 2 in A into alignment with its adjacent stator tooth as shown in (c). To move the slider further in the same direction, coil A is de-energized and coil B is excited in the opposite polarity to before. This is the state in (d).

The claw-tooth motor is another type of hybrid stepping motor. A cut-away view of this machine is illustrated in Fig. 3.35. Teeth are punched out of a circular steel sheet, and the circle is drawn into a bell shape. The teeth

Fig. 3.33 Disc-magnet stepping motor manufactured by Portescap

are then bent inside to form magnetic poles. The stator stack is formed by joining two bell-shaped casings so that the teeth of both are intermeshed and the solenoid coil is inside. The main benefit of a claw-tooth motor is its low manufacturing cost.

The claw-tooth motor was not accepted as a proper stepping motor when VR and hybrid motors were predominant, but it was used as an AC generator coupled to a motor shaft for detecting motor speed in feedback controls. As steel-sheet processing technology advanced, the positioning accuracy has improved to such an extent that it is now widely used as a stepping motor. Mass production of this type of motor has rapidly accelerated and now exceeds the number of hybrid motors.

3.5 DC motors

As discussed in Chapter 2, the DC motor is thought of as the basic motor, because it displays simple characteristics. Here, we will look at two aspects of the DC motor in order to understand its engineering problems. The first will be the roles of brushes and commutators to supply current to the rotor winding, and the second will be the methods of providing the magnetic field needed to produce torque.

Rotor (8) in Fig. 3.11 is the rotor of a DC motor. Note that this rotor is equipped with a 'commutator'. As shown in Fig. 3.36(a), the commutator

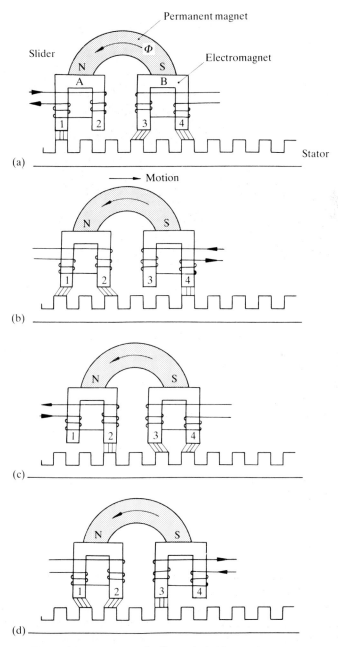

Fig. 3.34 Principle of a linear hybrid stepping motor.

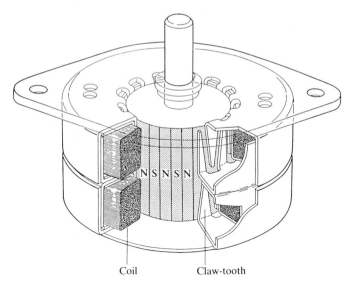

Coil   Claw-tooth

Fig. 3.35   Claw-tooth stepping motor.

consists of copper segments insulated from each other by an insulation material like mica or plastic. The number of commutator segments is basically the
same as that of slots, but must be at least three.

The DC rotor described in the previous chapter was a slotless rotor,
but most DC motors use a slotted core. For the latter, an odd number of
commutator segments is desirable, because even-numbered teeth increase
torque cogging due to symmetry with respect to the shaft. However, rotors
with even-numbered teeth (e.g. 12) are manufactured in large quantities,
because mechanized winding is easier with this shape.

The rotor is often referred to as an 'armature'. However, the connotations
of the two are different. A rotor is the rotating part of a motor, and the
armature is the part that carries the current interacting with the permanent-
magnet flux to create torque. In a DC motor, the rotor has armature coils
wound around it.

The connection between the commutator segments and one coil is illustrated in Fig. 3.36(b) for a nine-toothed core. Figure 3.37 shows the relationship between the nine coils, the commutator segments, and the brushes. The
nine coils are connected in series corresponding to the delta connection of a
three-phase AC motor. In terms of the AC motor concept, this is a nine-phase
motor, and a DC motor with three coils is a three-phase motor.

The DC current supplied from the positive terminal of the battery flows in
the winding through the positive brush, is divided into two paths, and collected at the negative brush to return to the minus terminal of the battery.

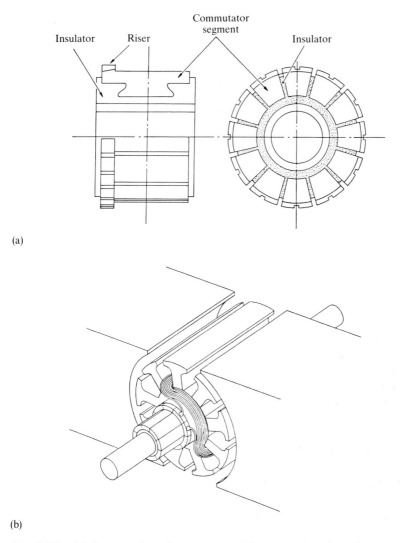

Insulator   Riser   Commutator segment   Insulator

(a)

(b)

Fig. 3.36 (a) Construction of commutator; (b) connection of a coil to commutator segments.

When the motor is revolving, the coils are moving, but the current paths are almost stationary. This was explained in the previous chapter using Fig. 2.9.

Next, let us consider the roles of the brushes more carefully in Fig. 3.38. This figure shows that one brush is equivalent to two switches placed in series. Recall the rotary switch used in British warships to operate the stepping

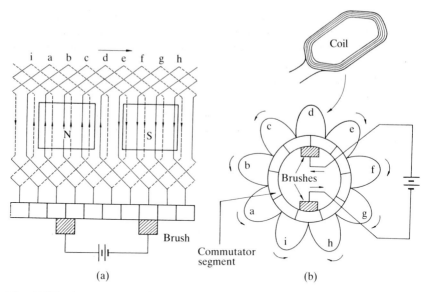

Fig. 3.37 Arrangement of coil, commutator segments, and brushes in a DC motor: (a) exploded diagram of lap winding; (b) coil connections.

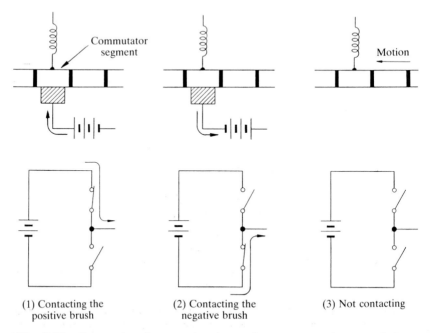

(1) Contacting the positive brush
(2) Contacting the negative brush
(3) Not contacting

Fig. 3.38 Connections between brushes and commutator via two switches.

motors in torpedo direction indicators (Fig. 2.5, p. 20). The function of the brushes and commutator is similar to that of the automatic rotary switch. When the rotor turns, the rotary switch automatically directs the current to the correct coils to produce the effective torque. Thus, the brushes and commutator are a clever assembly for making DC motors at reasonable cost.

An electromagnet can be used to provide field flux. The winding for the electromagnet is called the 'field winding'. There are basically three ways of connecting the field winding and armature through brushes which are supplied from a DC battery as shown in Fig. 3.39.

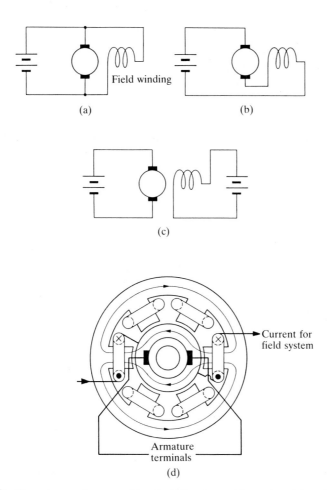

Fig. 3.39 Three types of connection of a DC motor with an electromagnet, using stator B shown in Fig. 3.3: (a) shunt connection; (b) series connection; (c) separate excitation; (d) field flux excitation and terminals for supplying armature.

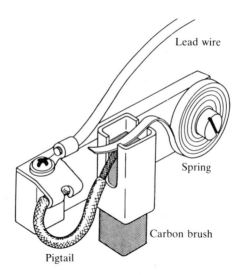

Lead wire

Spring

Carbon brush

Pigtail

Fig. 3.40   Brush holder used with stator B.

Scheme (a) is known as a shunt connection when the field and armature are parallel. With this connection, the motor turns at an almost constant speed if the applied voltage is fixed.

Scheme (b) is called a series connection. A high starting torque decreasing with speed is characteristic of this drive. If the motor does not carry a load, the speed will accelerate to a dangerous level. Traction motors are series DC motors.

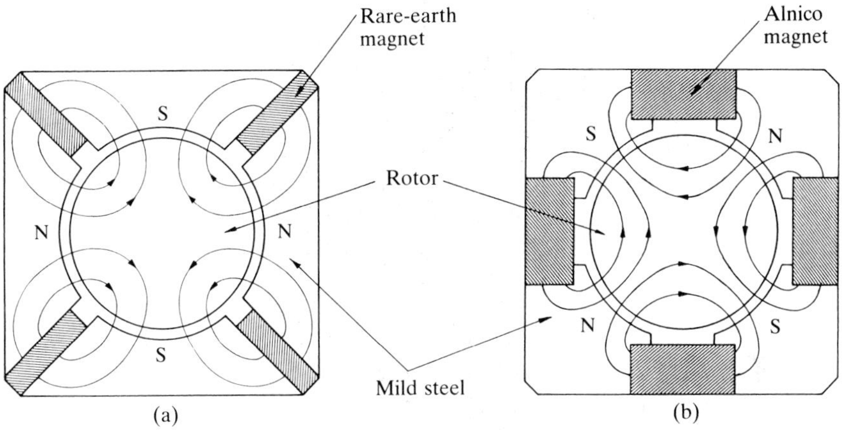

Fig. 3.41   Examples of arranging permanent magnets in DC motors: (a) using rare-earth magnets; (b) using Alnico magnets.

The driving method shown in scheme (c) is called 'separate excitation', because the field current and the armature current are adjustable separately to control motor speed.

Some basic experiments can be performed with the stator B in Fig. 3.3. Figure 3.39(d) shows how the field flux can be excited and how the power supply terminals are connected. Two of the six stator teeth are used for field poles here, but four poles can be used. Figure 3.40 shows a mechanism for holding the brushes.

Permanent magnets are used to a great extent in small DC motors. Figure 3.41 shows two examples of how to arrange permanent magnets. Another example using ferrite magnets is shown in Fig. 6.2 (p. 146). Thus the field system can be compact compared with an electromagnet DC motor, and no current is needed to provide the field flux.

The characteristics of a permanent-magnet motor are similar to those found in a separate-excited motor with a constant current in the field coil. The torque-versus-speed characteristics will be similar to those shown in Fig. 2.13 (p. 28).

The speed of a motor when it is not connected to anything is called the no-load speed. The no-load speed increases in proportion to the applied voltage. Let us look at a case in which the no-load speed is 50 rps at 5 V, and this decreases to 40 rps if a load of 0.1 N m is connected, as shown in Fig. 3.42. If the voltage is increased up to 8 V, the new no-load speed will be 80 rps, and the loaded speed 70 rps provided that the load is still 0.1 N m.

The most popular magnet used is made of ferrite, also known as a ceramic magnet. Though material cost is low, the available magnetic flux is relatively

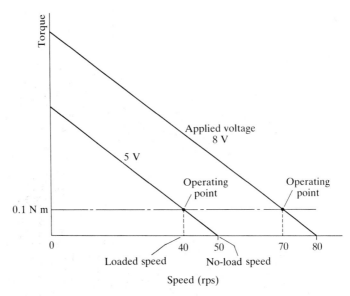

Fig. 3.42 Relationship between no-load speeds and loaded speeds in a DC motor.

low. For powerful motors, an expensive rare-earth magnet (e.g. samarium–cobalt or neodymium–iron–boron) is used. Alnico, an alloy of iron, aluminium, nickel, and cobalt, is also extensively used.

3.6 Universal motors

There is a motor similar to a DC motor but which can be operated with either direct current or single-phase alternating current. It is referred to as a universal motor, but is also known as an AC commutator motor or an AC series motor because the stator and rotor circuits are connected in series and designed to be operated on a single-phase 50/60 Hz supply.

The universal motor is powerful for its size, and turns at a very high speed when no load is carried. This motor finds application in power tools, such as small planers or sawing machines, food-processors, mixers, and vacuum cleaners. The drawbacks of this motor are noise, brush-wear, and sparking.

The biggest difference compared with a normal DC motor is that the stator core is replaced with laminated steel sheets. It is very dangerous to operate an electromagnet DC motor with a solid-steel core, since the core overheats because of the eddy currents induced in it.

Figure 3.43 shows a small power drill that uses an AC commutator motor.

Fig. 3.43 Universal motor used in a small power drill.

3.7 What is a brushless motor?

As explained in the previous chapter, the main features of DC motors are that they can be operated using a DC battery and the torque is proportional to the current. However, there is a drawback associated with the brushes and commutator. Because of the sliding contact between the brushes and the commutator, sparks are generated while the motor is running and both components can wear out. The brushes must be changed and the commutator resurfaced periodically. Small motors are often simply replaced. However, the reason that small DC motors are used in large quantities without changing brushes is that the lifespan of the instrument using them is usually shorter than the brush-wear time. For example, the DC motors driving the windscreen wipers of a car are only used occasionally.

However, in some applications such as expensive numerically controlled machines, brush wear is a serious problem that can only be solved by completely eliminating the brushes. As seen at the beginning of this chapter, AC motors of the revolving-field type, for example the squirrel-cage induction motor or permanent-magnet synchronous motor, do not have brushes or commutators. If these motors could be operated from a DC power supply, they would be brushless DC motors. When a permanent-magnet motor is used with an electronic inverter and position sensor, it can be a brushless DC motor with almost the same characteristics as a conventional one.

The electronic switching circuit used here replaces the brushes and commutator. We will see this using Figs 3.44 and 3.45. Figure 3.44(a) shows that one commutator segment performs the function of two series-connected switches. Now, each mechanical switch must be replaced with an electronic switch. A transistor is the simplest electronic device that can be used either as a signal amplifier or a switch. Its function as a switch is explained in Fig. 3.45(a).

However, we must take into account the adverse effect of sparks that occur when turning off a current. Once sparking happens in a transistor, its crystalline structure is dämaged, and it can no longer be used. A diode is very effective for eliminating sparks, as explained in Fig. 3.44(d). Diodes are tiny semiconductor elements like transistors, but function much more simply; a current can pass from anode to cathode, but is blocked in the opposite direction (see Fig. 3.45(b)).

As shown in Fig. 3.44(e), two transistors and two diodes are needed to replace a commutator segment. In practice, additional transistors and some logic elements are needed in the preceding stages for processing switching signals coming from a rotor-position detecting device.

If a DC motor with nine commutator segments is to be transistorized, the cost of the electronics will be high. The minimum is three and hence a three-phase switching circuit is often used for brushless DC motors.

The structure of a brushless DC motor is similar to that of an AC synchronous motor. In other words, a brushless DC motor is a permanent-magnet motor driven with a transistorized mechanism for brushes and commutator to provide it with DC motor characteristics.

Let us look once again at Fig. 3.24. By using three switches to control the current directions in the windings, one can make the magnetic field revolve. When a permanent magnet is placed in this magnetic field, the rotor will revolve following the field movement. However, this method has the following disadvantages.

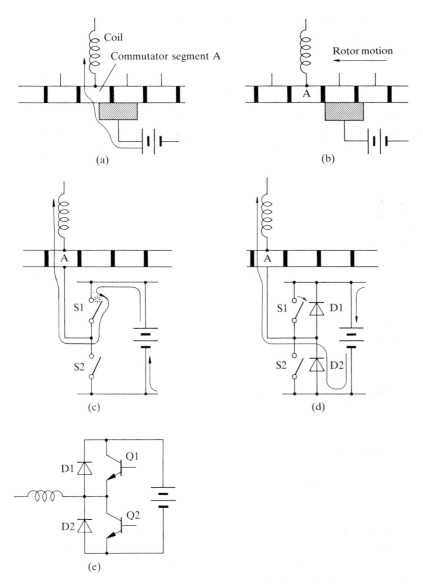

Fig. 3.44 The function of brushes and commutator is replaced with transistor circuits. (a) Commutator segment A is in contact with a brush; (b) when it separates from the brush, the remaining current tends to flow through air breaking insulation: this is sparking; (c) this is simulated by a circuit with two switches S1 and S2; (d) if a diode is placed in parallel with each switch, however, a transient current is supplied from the negative terminal of the power supply and no sparks occur in S1; (e) each switch is replaced with a transistor to form an electronic circuit.

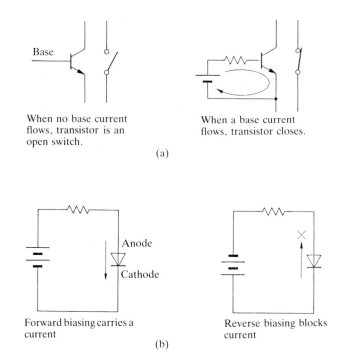

When no base current
flows, transistor is an
open switch.

When a base current
flows, transistor closes.

(a)

Anode

Cathode

Forward biasing carries a
current

Reverse biasing blocks
current

(b)

Fig. 3.45 (a) Transistor as an electronic switch; (b) basic function of a rectifier diode.

1. If the switching frequency is too high, the rotor will not start owing to its large inertia.

2. Since the rotor speed is expected to be the same as the rotating speed of the magnetic field, speed control seems easy. In fact, however, a low-frequency speed fluctuation often occurs, and is a very complicated phenomenon to suppress.

To overcome these difficulties, the timing of switching should be automatically determined from the rotor position. In such a scheme, the rotor will always start. The second problem is also solved by adding the proper control circuit, to be discussed below.

Let us look at the system shown in Fig. 3.46, which uses electronic techniques for detecting the position and generating the switching signals. Six phototransistors are placed on the end-plate at equal intervals. Since a shutter is coupled to the rotor shaft, these photoelements are exposed in sequence to the light emitted from a lamp situated at the left of the figure.

Now the problem is the relationship between the ON/OFF state of the transistors and light-detecting phototransistors. The simplest configuration is

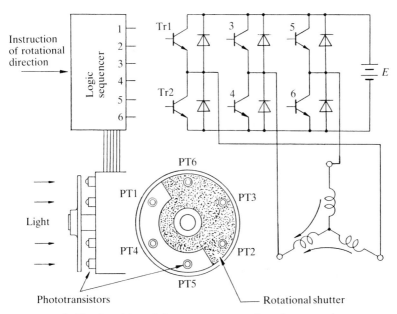

Fig. 3.46 Brushless DC motor system using phototransistors.

when the logic sequencer is arranged in such a way that when a photo-transistor marked with a certain number is exposed to the light, the transistor of the same number switches ON. In the state of Fig. 3.46, transistors 1, 4, and 5 are ON. The rest of the transistors are now OFF.

Now, if the rotor is arranged so that the relationship between the stator magnetic field and the rotor position is as shown in state (1) in Fig. 3.47, a clockwise torque will act on the rotor. After it revolves through 30°, PT5 is switched OFF and PT6 ON, which makes the stator's magnetic pole revolve 60° clockwise. This is state (2), which causes the rotor to revolve further in the clockwise sense. After turning another 60°, the transistors 3 and 4 change ON/OFF states, which makes the stator field revolve a further 60° clockwise to get to state (3), and this further proceeds to state (4). Accordingly, the rotor will receive a clockwise torque at any position.

By replacing the brushes and commutator with an electronic circuit, the torque-versus-speed characteristics become very similar to those obtained with a DC motor. The shutter shown is asymmetrical and can cause mechanical vibration. However, because most brushless DC motors are of four or more poles, the shutter will be symmetrical.

Let us discuss how to reverse this motor. A conventional DC motor can be reversed by simply exchanging the terminal polarities connected to the DC supply. However, in this case, this is not possible because transistors are

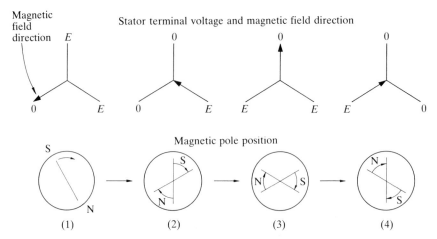

Fig. 3.47 Magnetic rotation accompanied by rotor rotation.

unidirectional switches and would be damaged by such an operation. Another method is to turn the shutter position through 180° to reverse the ON/OFF relationships of the transistors, but this is still not practical.

A practical method of reversing the rotational direction is to arrange the logic sequencer in such a way that when a photodetector marked with a certain number is exposed to light, the transistor of the same number is turned OFF. Conversely, when a photodetector is not exposed to light, the transistor of the same number is turned ON. By this arrangement, the currents in the three phases are all reversed and the torque direction changes. Such a design for the logic sequencer presents no difficulties.

A very popular method of detecting rotor position is the use of a semiconductor device called a Hall-effect sensor. A typical Hall-effect device is a small pellet of semiconductor material in which electrons are charge carriers.

The principles of the position sensor using a Hall-effect element are shown in Fig. 3.48. It is necessary that a current I is always flowing in the pellet used as a flux detector. This magnetic flux may be coming from the rotor's permanent magnet itself or another permanent magnet coupled to the motor shaft. If the pellet is exposed to the flux, as depicted in this figure, the electrons moving downwards are directed to the left-hand side in accordance with Fleming's left-hand rule, which results in negative polarization on this side and positive polarization on the other side.

Thus, by finding a biased potential of this polarity across the output, the pellet is used to detect a north pole. If the magnetic field alternates owing to the rotor's rotation, the charge carriers in the pellet will be directed to the other side, yielding a reversed output voltage.

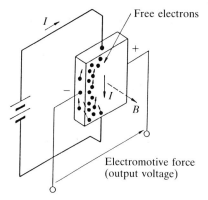

Fig. 3.48 Principle of Hall-effect element: electrons are directed to the side in a magnetic field when a current is flowing.

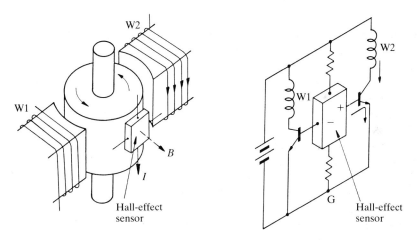

Fig. 3.49 Simplest brushless DC motor with a Hall-element device.

This phenomenon, known as the Hall effect, was discovered by an American scientist E. H. Hall in 1878 from an experiment with a piece of metal. The Hall effect is stronger in a semiconductor than in a metal.

A way of detecting the rotor position and generating a switching signal is as follows. Figure 3.49(a) shows a simple set-up consisting of two stator poles with a coil at each, a permanent-magnet rotor, and Hall-effect element placed close to the rotor surface. Figure 3.49(b) shows the simplest connection between the Hall-effect element and the two transistors governing the currents in coils Wl and W2.

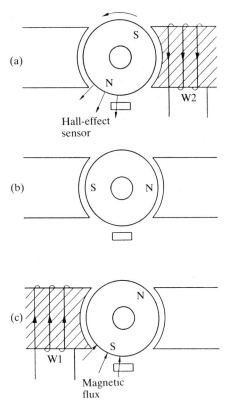

Fig. 3.50 How switching is carried out in reference to the rotor position using a Hall-element sensor in the configuration shown in Fig. 3.49.

Figure 3.50 explains how the rotor continues to turn, as follows.

(a) In this state, the Hall-effect element detects the rotor's north pole, and coil W2 is energized to produce the south pole which drives the rotor in the counterclockwise direction.

(b) Since no magnetic field is applied to the Hall-effect element in this position, both transistors are in the OFF state, and no currents flow in Wl or W2. The rotor will continue to revolve because of its inertia.

(c) The Hall-effect element detects the rotor's south pole, and winding Wl is energized to create a south pole·which attracts the rotor's north pole to continue the counterclockwise motion. By repeating this sequence the rotor will turn continuously.

A fault with such a simple scheme is that the motion may stop in state (b) if a rotor is carrying a heavy load. Therefore, most brushless motors use two

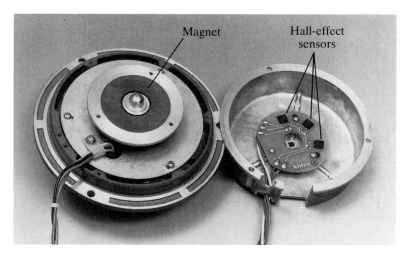

Magnet   Hall-effect
                sensors

Fig. 3.51   Inside a brushless DC motor with a four-pole configuration. The magnet exposing three Hall-element devices is separate from the rotor.

or more Hall-effect elements and three or four windings so that an effective torque is produced at any rotor position. Figure 3.51 shows a brushless DC motor equipped with an external magnet to expose the Hall devices. Here, three Hall elements are arranged at 60° intervals.

A brushless DC motor is similar in structure to a permanent-magnet synchronous motor, a sort of AC motor. However, as pointed out above, brushless DC motors are equipped with rotor-position detectors and mechanisms to feed back the position information to the switching circuit.

On the other hand, a brushless DC motor displays behaviour similar to that of a conventional DC motor. The most characteristic feature of a DC motor is that the torque is proportional to the current supplied, as discussed in reference to equation (2.4) in Chapter 2. In other words, if a torque of 1 N m is produced by a current of 1 A, a torque of 2 N m occurs with 2 A, independent of speed.

Also, if the relationship between the torque and rotational speed is plotted on a graph, we obtain a set of parallel straight lines with decreasing slopes. Again, the slope is independent of the applied voltage or speed. A brushless DC motor shows similar characteristics.

Next, let us look at the similarities and differences from the so-called commutation point of view. First, in respect of the machine structure, there is a big difference, as follows.

- Conventional DC motor: the armature is on the rotor, and the field system contains either an electromagnet or a pair of permanent magnets attached to the stator core.

● Brushless DC motor: the stator carries the armature, and the rotor carries magnets supplying the field flux.

The fundamental reason for such differences comes from the basic difference in commutation methods. Here, 'commutation' means sequential switching of the supplied DC current to the coils of the armature. In an orthodox brushless DC motor, the windings are connected in a three-phase star or delta scheme, and commutation is implemented by an electronic circuit as shown in Fig. 3.24. If we try to construct the complete electronic equivalent of the commutator mounted on the rotor, a lot of transistors are needed, as indicated in Fig. 3.44. Carbon brushes and a copper commutator play the role of an expensive and sophisticated electronic circuit. In this sense, we can say that the combination of brushes and commutator is a cheap, clever device.

It is obvious that an increased number of phases in a brushless DC motor, five or seven for instance, is not realistic in terms of cost. This is why most brushless DC motors are three-phase. However, there are some four-phase and economic single-phase ones, as well.

Conclusions

As we have seen in this chapter, motor technology involves machine construction, materials, electronics, sensors, and control technology, and they are all developing rapidly. In this chapter, we have studied basic stator and rotor structures, and commutation technologies. Current topics in machine construction and other aspects will be dealt with in Chapter 6. In the next chapter, we will survey speed and position-control techniques of motors.

4
Classical motor-control technology

Aeroplanes are driven by reciprocal or jet engines, and most automobiles by petrol or diesel engines. Electric vehicles are at present very few, though trains are normally driven by electric motors. Most large factory machines, small home appliances, and business machines are driven with motors as the prime mover. Hydraulic and pneumatic systems are also used for special purposes. Electric motors are known to be easier to control as far as speed and rotational angle go than other prime movers. In this chapter, we will see how to control motor speeds and rotational angles by referring to several simple examples.

4.1 Dependence of AC motor speed on frequency

When motor specifications are determined in the design stages, the speed or speed range and how to control that speed are taken into account. As the problems with an AC motor are rather simple, we will start with it. In the UK, the electric power supplied to homes is 220 V with a frequency of 50 Hz. When a typical four-pole AC motor is driven on this net, the machine will rotate at 25 rps or a little slower. Home-appliance devices are designed to run well when the motor turns at this speed.

The same is true with factory motors. Therefore, if a certain type of motor were to be run under different circumstances, there would be a problem. Here is an example: in a technology-aid project, a French-made machine was installed at a technical school in a developing country in Asia. The machine was designed for operation at 50 Hz, but the frequency of that country is 60 Hz because of American influence. The motor ran at 20% higher than normal speed, which was very dangerous for that machine and its operators.

A teacher from this school had a chance to go to Japan to study modern technology, expecting to find some clever method to solve this problem. However, he found it to be rather annoying. It was suggested that, if he could build a device called an inverter, the motor speed could be reduced. That was before inverters began to be manufactured at a reasonable cost in advanced countries. Inverters are very popular now for easy speed control. Before studying modern inverter technology in the next chapter, let us examine some classic techniques of speed adjustment and the basic principles of the inverter.

4.2 Classical method of changing speed with a single-phase AC motor

Single-phase AC power is available in every house. This is why home appliances normally use a single-phase AC motor. Before electronic controls became very popular, the constant speed of the magnetic tape in a tape recorder or the turntable of a record player had been achieved with a synchronous AC motor. The rotational speed of a synchronous motor is proportional to the power supply frequency and inversely proportional to the number of magnetic poles.

The capstan, which is a cylindrical rod holding the magnetic tape sandwiched with a rubber roller, is driven at two or three different rotational speeds in an open-reel tape recorder. For this purpose, two or three sets of windings with different numbers of poles are installed in the stator. When the four-pole winding is used the speed is 25 rps to drive the magnetic tape at 19 cm s^{-1}, and if the eight-pole windings are used the speed is 12.5 rps for 9.5 cm s^{-1} tape

Table 4.1 Relationship between number of poles, frequencies, and speeds in AC synchronous motors

Number of poles	Speed at 60 Hz (rps)	Speed at 50 Hz (rps)
2	60	50
4	30	25
6	20	16.6 ...
8	15	12.5

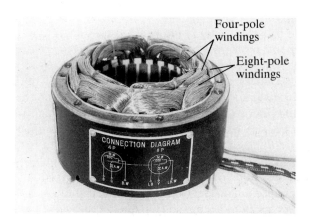

Four-pole windings

Eight-pole windings

CONNECTION DIAGRAM

Fig. 4.1 Stator of a hysteresis motor with four-pole and eight-pole windings used in a tape recorder in broadcasting stations.

speed. A two-pole configuration produces a speed of 50 rps for 38 cm s^{-1} for the same capstan size.

The relationship between the number of poles, frequency, and speed is given in Table 4.1. In practice, three-speed drive is rare, but two-speed motors with four- and eight-pole windings have been manufactured in large numbers. Figure 4.1 shows the stator of a motor designed for a tape recorder used in a broadcasting station.

In a more sophisticated configuration, coil connections are changed by switches to produce two different numbers of poles (e.g. 2/4, 4/6, etc.). Pole amplitude modulation or PAM was designed by the British inventor G. H. Rawcliffe, and was suited to squirrel-cage induction motors for power use.

4.3 Variable-speed drive with controllable frequencies

The disadvantages of changing speeds by selecting the number of poles is that the speed ratios are limited to integral ratios such as 1 : 2, 1 : 4, or 2 : 3, because the number of poles is always an integer and definitely an even number.

If we can vary frequency continuously, we have an ideal variable-speed drive. An inverter produces variable-frequency AC current. Literally, the inverter is a device that converts DC power to AC power using transistors or similar solid-state devices. However, in many practical inverters, the DC power is provided by a three-phase commercial network. Figure 4.2(a) shows a block diagram of an inverter. The stage of converting the input AC power to DC consists of diodes and is called a rectifier. The inverter or the stage converting the DC back to AC has the same configuration as the transistor circuit for the brushless DC motor discussed in the previous chapter (see Fig. 3.46). Figure 4.2(b) shows a simple, laboratory-assembled inverter using bipolar transistors. (The basic functions of the transistors will be discussed in Chapter 5.) This inverter is so simple that it is only suited to a small motor. The switching sequence used in this inverter is that shown in Fig. 3.25, and hence the output voltage is not variable. In sophisticated modern inverters not only the frequency but voltage is adjustable, owing to the pulse-width modulation technique that will be discussed in the next chapter.

Recently, inverters have been widely used from low power to quite high power. One of the major application areas is the control of air flow in a building. The use of an inverter is the best way of solving the inconvenience that occurred in the foreign-aid story above. It is, however, very difficult for a technical school teacher to build a reliable inverter by simply assembling components. The easiest way would be to request the French Government to provide an appropriate inverter made in France. Inverters are also used for the high fixed speed of the polygonal mirror in a laser printer (see Fig. 1.4).

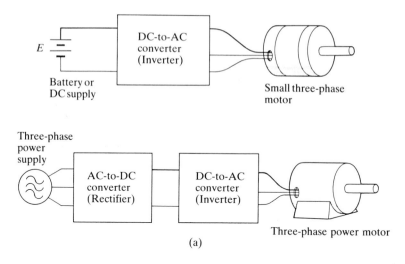

(a)

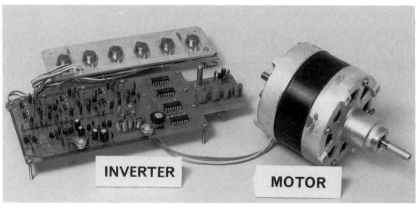

(b)

Fig. 4.2 (a) System diagrams of inverters; (b) a simple practical inverter for turning a small motor at a fixed speed independent of power-supply frequencies.

4.4 Simple method of adjusting single-phase AC motor speed

Electric fans are normally driven by an AC motor. When adjusting the fan speed, is an inverter always used? The answer is, not always. The large fans for air flows in a building or factory are controlled by an inverter. However, an inverter is not suitable for the speed control of single-phase AC motors.

A simple and low-cost method makes use of the conduction-angle control technique. Look at Fig. 4.3. The waveforms in portions (a) and (b) illustrate this concept; the whole voltage is not applied to the motor. In (a) the supply

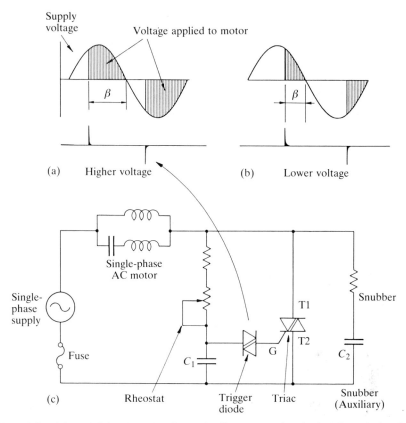

Fig. 4.3   (a) and (b) Principle of speed adjustment of a single-phase induction motor by means of conduction-angle control; the conduction period is controlled by an electronic device called a triac. (c) A simple example of a practical circuit.

voltage is applied at a timing of about 50° from the start of a half-cycle. The period in which the voltage is applied to the motor is called a 'conduction angle' and denoted by β; it is 130° in (a). The conduction angle can be varied from 180° to zero. (Here one cycle of the alternating voltage is regarded as a 360° interval.) In the case (b), the conduction angle is about 50°. The larger the conduction angle, the faster the motor turns.

To carry out this control, a solid-state device called a triac is suitable, and it is used with a trigger diode, capacitors, and a small rheostat in a circuit arrangement shown in Fig. 4.3(c). The basic functions of the triac and trigger diode will be briefly explained in the next chapter. Here, capacitor C_1 and the trigger diode are essential to generate the trigger signal with a time lag behind the start of a half cycle of the power supply. The timing of the trigger signal, and hence the conduction angle, can be adjusted by the rheostat.

Note that a capacitor is used in series with one of the sets of the motor windings. As was shown in Fig. 3.10 on p. 59, when a small three-phase motor is operated on a single-phase network, an appropriate capacitor is connected in parallel with one of the three coils. However, in the case of a motor with two-phase windings, a capacitor is normally placed in series with one coil. The capacitor and the winding together generate an imperfect four-phase (therefore two-phase) current from a single-phase current.

Figure 4.4 illustrates what the control set-up will be like. Details of this technique are described in one of the author's books[1] for college level.

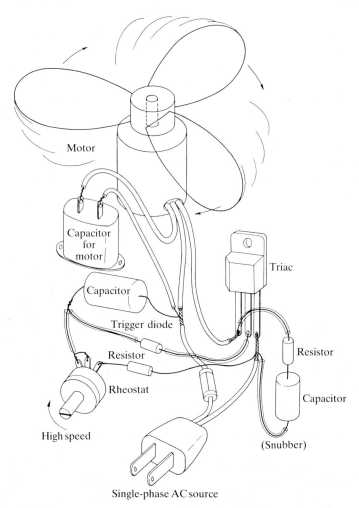

Fig. 4.4   Component arrangement of the circuit in the previous figure.

This speed-adjustment method is very simple, but not always suitable for applications where high-energy-efficiency drives are essential, because it is, in principle, not desirable to vary the speed of an AC motor by changing the applied voltages. As explained in the previous chapters, the induction and synchronous motors are the most important members in the category of AC motors. The speed of a synchronous motor is absolutely determined by the number of poles and the frequency, but voltage is not the factor determining speed.

4.5 Conventional DC motors regulated by voltage control

As explained in Chapters 2 and 3, a conventional DC motor rotates simply by connecting its two terminals to a battery. Unlike AC motors, the voltage is the only factor to affect the speed, because the frequency of DC power is absolutely zero. (Recall that DC current is converted into AC current in a DC motor by means of commutator and brushes.) The motor speed at no load is proportional to the voltage applied to the terminals. However, in practice, it is impossible to control the speed only by adjusting the voltage when the motor carries a load. The speed decreases with load. Therefore, for fine adjustment of speed, the speed or the load quantity must be measured, and the voltage adjusted from that information. Now, let us compare three typical methods.

4.5.1 Mechanical governor method

This method, in which the speed is measured by centrifugal force, has been used for a long time (see Fig. 4.5). When the speed is above a certain level, a mechanical contact that rotates with the rotor assembly separates and the armature circuit is opened. In short, the applied voltage becomes zero and the speed decreases. When the speed drops below a certain level, the mechanical contact closes and a voltage is applied to the armature circuit again. The speed then increases, the contact separates, and so on. Under steady conditions, the contact opens and closes with high frequency at a particular motor speed. Thus, the average voltage is automatically adjusted and this corresponds to a particular motor speed.

4.5.2 Speed control by means of an electronic governor

The mechanical governor has declined in popularity, and the electronic governor is becoming more important. A DC motor can be a DC generator as well. Even when it is run as a motor, the generator action exists and the voltage generated is proportional to the running speed. This generated voltage

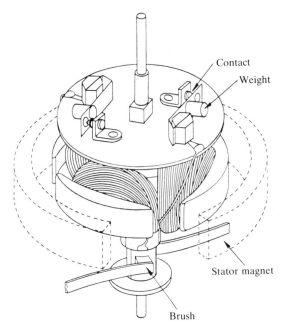

Fig. 4.5   DC motor with a mechanical governor for automatic speed adjustment.

can be read by an electronic circuit and utilized as feedback for the driving circuit to adjust the applied voltage so that the speed is controlled: this is an electronic governor drive.

Figure 4.6 shows an example of a DC motor that has an electronic governor. Even with this configuration, however, a high accuracy of speed control cannot be expected because of the undesirable effect of brush-voltage drops and temperature-dependent winding parameters.

An electronic governor is applied in a brushless DC motor that uses transistors in place of brushes and a commutator.

4.5.3 Speed control using a tachogenerator

For a higher grade of speed control, the use of a proper speed sensor is needed. A DC tachogenerator is an analog device that is sensitive even at very low speeds. Figure 4.7 shows a DC motor equipped with a tachogenerator. A tachogenerator's construction is very similar to a conventional DC motor, but it is designed to generate an output voltage proportional to the rotating speed. In this example, both the motor and tachogenerator have a so-called moving-coil rotor, which is very light and quick at changing speed.

The principle of the speed control using a tachogenerator can be best

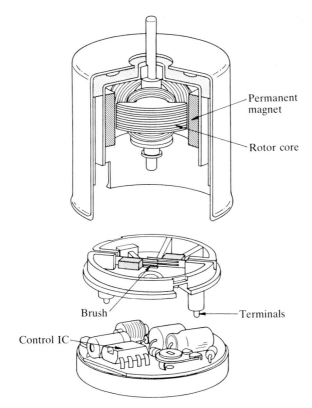

Fig. 4.6 DC motor with a built-in electronic governor.

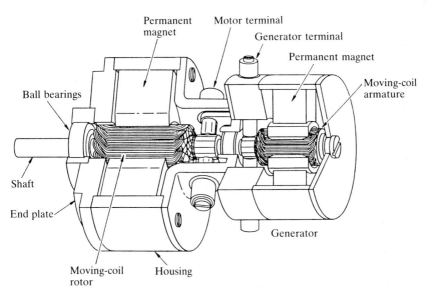

Fig. 4.7 DC motor mounted with a tachogenerator.

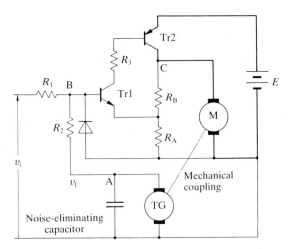

Fig. 4.8   A speed-control circuit using a tachogenerator.

explained by the simplified circuit shown in Fig. 4.8. Since the practical circuit is more complicated, the transistors here are ideal transistors and only the important parts are illustrated. The principles are as follows.

1. The speed instruction is given by the voltage: here it is 5 V.

2. The tachogenerator is directly coupled to the motor shaft, and its terminals are connected so that a negative voltage proportional to the speed appears at terminal A. Here the tachogenerator is supposed to provide an output of 5 V at 20 rps.

3. Resistors R_1 and R_2 are equal, and at point B the voltage is half of the difference between the applied voltage v_i and the tachogenerator voltage v_f, that is, $\frac{1}{2}(v_i - v_f)$. Therefore, when revolving at 20 rps, the voltage at point B must be zero. Now, the speed is lower than this, and hence a positive voltage will appear.

4. Transistor Tr1 and resistors R_A and R_B make up a voltage amplifier, and its gain is $(R_A + R_B)/R_A$. For instance, if R_B is 90 Ω and R_A, is 10 Ω, the gain is 10. The amplified voltage appears at point C.

5. Transistor Tr2 is for power amplification. In other words, it is a transistor that draws out the current needed by the motor from power supply E.

6. Consider the case when the motor is running with a voltage of 4 V at point A. Then, the command voltage (5 V) minus this voltage is 1 V. Hence 0.5 V will appear at point B. It is amplified to 5 V by a factor of 10, and applied to the motor. The motor will be accelerated by this voltage. As a result, the

voltage at point A is boosted, and the voltage at point C drops. The motor will settle to an appropriate speed.

7. Eventually, the voltage at point A will become a little lower than 5 V, for instance 4.8 V, and the speed will be maintained, about 5 V being applied to the motor.

Although the analog controller is simple, this technique has not been used widely because the speed regulation is not accurate owing to temperature-dependent resistance and age-varying capacitance in the control-stage components. Furthermore, the tachogenerator is an expensive component.

4.6 Speed control using a photo or magnetic encoder

For high-accuracy speed control, a pulse-generator sensor is needed. One typical type is a photoencoder, also known as an optical encoder.

4.6.1 Photoencoder

There are basically two fundamental types of photoencoders. One is the simple one shown in Fig. 4.9, where a disc with slits in it is coupled to the rotor shaft and rotates near a fixed slit attached to the stator. While rotating, the fixed slits pass and block the light from a light source. A photosensitive semiconductor device receives the interrupted light signals and converts them to electrical pulses. The pulse frequency is proportional to the rotor speed.

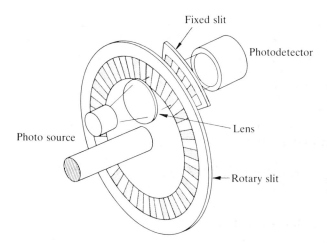

Fig. 4.9 Principle of an incremental photoencoder.

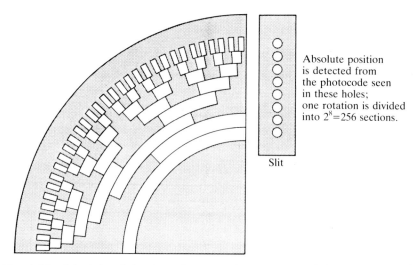

Absolute position
is detected from
the photocode seen
in these holes;
one rotation is divided
into $2^8=256$ sections.

Slit

Fig. 4.10   Principle of absolute photoencoder; a quarter of the disc pattern is
shown. By increasing channels, the angular resolution is increased.

This type of device is called an 'encoder', where the speed information is
encoded in the form of a pulse train.

Another type of encoder is a device that converts the rotor position into a
binary code with several digits. The rotating disc has a pattern, as shown in
Fig. 4.10, and the fixed unit has only one slit but multichannels. This is known
as an 'absolute encoder'.

Using a simple incremental method, it is impossible to detect the rotational
direction and the absolute position. However, by taking the signals from two
appropriate locations, the rotational direction can be determined, and, by
counting the pulses, the position can also be calculated. For this, some
electronic logic circuit or a microprocessor is needed. Magnetic encoders
have recently become popular .

4.6.2 Magnetic pulse generator

The principle behind the speed sensor widely used in record players and floppy
disk drives is shown in Fig. 4.11. This is a combination of fine magnetization
in the rotating member and a printed circuit with a zigzag pattern. From
Fleming's right-hand rule, the induced voltage in the printed circuit is an
alternating voltage, and its frequency is proportional to the speed.

4.6.3 Using an incremental encoder

We shall examine two methods of using an incremental encoder to control a
motor's speed.

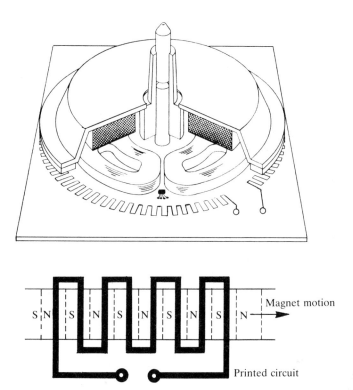

Fig. 4.11 Principle of a speed sensor using high-resolution magnetization and a printed zigzag circuit. The frequency generated in the printed circuit is proportional to the speed.

1. *Frequency method.* Here, speed is controlled with an analog voltage. The pulse signal from the sensor is inputted to a circuit called frequency to voltage (F/V) converter, and the output voltage is proportional to the pulse frequency. The other parts are very similar to those of the configuration using a tacho-generator. Figure 4.12 should be referred to for the principle of the F/V converter. Despite the digital pulse train used here, this method is analog, and the speed-control accuracy is low.

2. *Phase-locked servo* (*PLS*). This is a method giving instruction with pulses, as illustrated in Fig. 4.13. The speed instruction is given by a pulse train. The control is carried out so that the encoder pulse train is synchronized with the command pulse train. In other words, the phase difference between the command pulse train and the feedback pulse train is locked in this control scheme.

The greatest advantage of the phase-locked method is that the speed never varies with the motor load. This type of control is used in audio and video equipment, high-quality laser printers, etc.

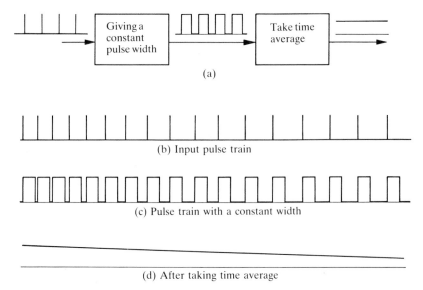

Giving a constant pulse width
Take time average
(a)
(b) Input pulse train
(c) Pulse train with a constant width
(d) After taking time average

Fig. 4.12 Principle of a pulse F/V converter.

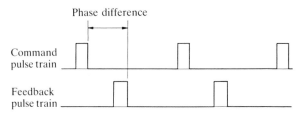

Phase difference
Command pulse train
Feedback pulse train

Fig. 4.13 Principle of the phase-locked servo: the phase difference between the command pulse train and the feedback pulse train is maintained.

4.7 Servoamplifier for a bidirectional DC motor drive

Above, we saw an example of speed control for a DC motor using a transistor circuit and a tachogenerator. The transistor circuit discussed was only for driving the motor in one direction; reverse operation is impossible because the voltage applied to the motor is always either positive or zero and a negative potential is not available from this circuit.

An example of a bidirectional circuit is shown in Fig. 4.14, using resistors, diodes, and two different types of bipolar transistors: *npn* and *pnp*. Fundamental properties of transistors are dealt with in Chapter 5, and details of these devices are not essential here. The function of each part is more important. The input voltage v_i, which may be either positive or negative but within

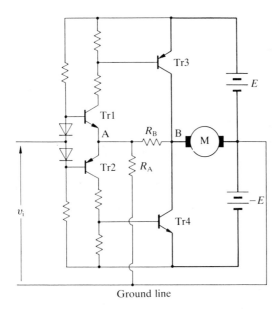

Fig. 4.14 An example of a bidirectional voltage-controlled servoamplifier using bipolar transistors.

E and $-E$, appears at point A. Resistors R_A and R_B are very important because they act as a voltage amplifier with gain $(R_A + R_B)/R_A$. If, as before, R_A is $10\,\Omega$ and R_B is $90\,\Omega$, the gain is 10, and a voltage of 10 times v_i appears at point B, which is the voltage applied to the motor terminal. Transistors Tr3 and Tr4 form, in technical terms, a power amplifier. In other words, the current needed by the motor is supplied from power supply $+E$ or $-E$. When the voltage at point B is positive the current is supplied from $+E$ through Tr3, and, if it is negative, Tr4 draws a negative current from $-E$. Such a transistor circuit is called a servoamplifier. This servoamplifier is classified as a voltage-control servoamplifier because what it controls is the voltage applied to the motor. Current is determined by the voltage and the motor speed.

4.8 From voltage to current control

When the load changes or a noise develops in the electronic circuit, the motor speed may go up or down. A method of restoring the speed is by voltage adjustment in a feedback-control system, explained above. However, this is not widely used in practical systems because the current variation always lags behind the voltage in an electric circuit driving the motor.

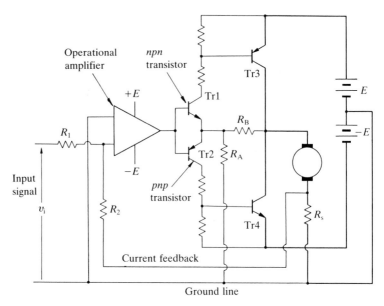

Fig. 4.15 An example of a bidirectional current-controlled servoamplifier.

A more desirable and prevailing method is a current-control configuration, though its main drawback is its somewhat high cost. In motion control, it is most effective for controlling the torque, since it directly affects acceleration or deceleration. Since, in a DC motor, the torque is proportional to the current, controlling the current means controlling the motor torque.

An example of a current-controlling servoamplifier is shown in Fig. 4.15. It can be seen that a device called an operational amplifier is added. Resistor R_s, which is for sensing the motor current, is added too. In this configuration, the motor current is determined by the input voltage v_i by the following equation:

$$i_m = -(R_2/R_1 R_s)v_i. \tag{4.1}$$

The motor voltage is automatically determined by the current and motor speed.

4.9 Controlling a stepping motor with pulse signals

Stepping motors are similar to AC motors in structure. However, they are regarded as belonging to a completely different category. Unlike an AC or a DC motor that runs when its terminals are connected to a power supply, a stepping motor is always driven with the help of an electronic circuit. As explained in Fig. 4.16(a), electric power is supplied to the drive circuit, which normally uses transistors, from a DC supply.

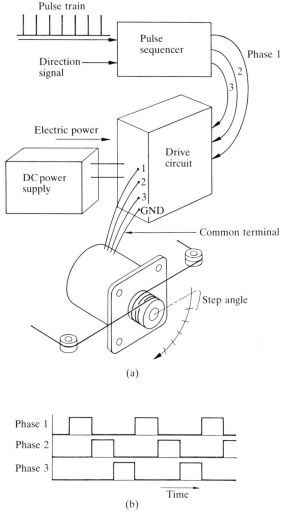

Fig. 4.16 (a) Drive system of a stepping motor; (b) switching sequence for a three-phase scheme.

In this example, the stepping motor has three phases or three pairs of windings, and initially a current is flowing in the first phase windings as shown in Fig. 4.16(b). In this state, this current yields a holding torque, and the motor would not move even if an external torque is applied to the motor shaft, as long as it is below a certain level.

Then, when a pulse is applied to the pulse sequencer which is located in front of the drive circuit, the current is switched from the first phase to the second phase. In technical terms, a 'commutation' takes place, and this causes the motor to step a certain angle, known as the 'step angle'. The motor will exert a holding torque at the new position to resist an external torque. If another pulse is applied, the current is switched to the third phase to rotate the motor through one step angle. Therefore, if pulses are applied continuously at a constant rate, the motor will rotate at a constant speed.

When the pulse rate is very low, the motion will be step-like. However, when the pulse frequency is higher, the movement becomes smooth.

If a pulse train of 30 pulses is applied, the motor will rotate through 30 step angles and exert the holding torque at the final position. Thus, a stepping motor can be used either for speed control or for position control.

When reversing the rotational direction, the signal given to the direction terminal of the logic sequencer must be reversed. For example, if the signal level is high, normally 5 V, and the direction is clockwise, then a counter-clockwise motion will result from changing the signal to a low level, normally zero volts.

A big advantage of the stepping motor is that it can control both rotational direction and the position at which the motor stops, and can generate a strong holding torque from a simple configuration using digital signals. Typical values of the step angle are 15°, 7.5°, 2°, 1.8°, and 0.9°. The stepping motor finds extensive application in magnetic-head positioning in floppy and hard disk drives. The daisy wheel in a printer or an automatic typewriter is also driven by a stepping motor.

4.10 Classical means of position control

There are many applications of motors in position control. The most well-known area is the control of a robot's arm. However, a more simple example is the control of the rotational angle of a disc or a pulley coupled to the motor shaft. Simply speaking, this is controlling the rotational angle of the motor. We will see here a simple classical position-control method using a DC motor.

A test set-up of this control system is illustrated in Fig. 4.17, where a voltage-controlling servoamplifier shown in Fig. 4.14 (p. 111) is used. The most characteristic feature of this system is that a position sensor known as a potentiometer is coupled to the motor shaft. Figure 4.18 illustrates a potentiometer and indicates its function. Position information is available as a voltage output proportional to the rotational angle from the reference position where the sliding contact comes to the negative terminal of the applied voltage. In addition to the position sensor, a tachogenerator is also used to produce a speed feedback voltage in this positioning system.

In this system, too, the position command is given by a voltage. The voltage difference between the command voltage and the output voltage from the potentiometer is taken and amplified in the servoamplifier, to be applied to the motor. Let us see what will happen next.

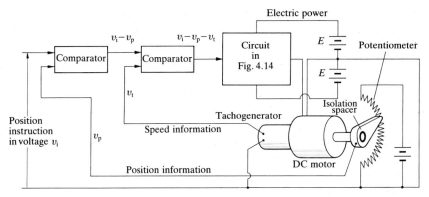

Fig. 4.17 Classical position-control system using a potentiometer as the position sensor. A tachogenerator is also used to stabilize the motion.

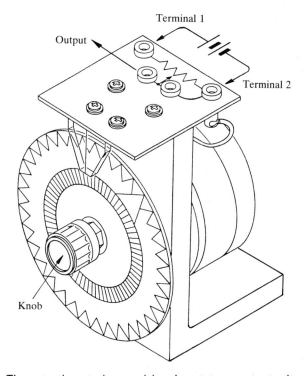

Fig. 4.18 The potentiometer is a precision rheostat: a constant voltage is applied across terminals 1 and 2, and the output is taken from the sliding contact. This potentiometer can be used for giving position instructions in the set-up shown in the previous figure; a voltage proportional to the knob angle is generated from the output terminal. If a motor shaft is coupled instead of the knob, the potentiometer is an angle sensor.

At first, we assume the position command to be 5 V, and the potentiometer output voltage to be 3 V. Therefore the voltage difference is 2 V. As the motor is not moving at first, the speed feedback voltage is zero, and hence this 2 V is not affected. We assume that the amplifier's gain is 10 times. The voltage difference is amplified to 20 V, and it will be applied to the motor. The motor will begin to rotate in the direction which makes the feedback or potentiometer's output voltage higher. When it becomes 4 V, the voltage difference is 1 V and motor voltage is 10 V. It is likely that the potentiometer's output will be automatically controlled to become 5 V and the motor positioned at this point.

However, if the tachogenerator were not used here, there would be a problem. Since the motor does not reduce speed as it approaches the target position, it would overshoot. For example, the motor may stop where the output is 7 V. The voltage difference is now -2 V, and this is amplified to be -20 V. The motor will soon be accelerated in the opposite direction toward the target. It is obvious that it will overshoot again. Thus, oscillation would occur around the target position.

Something is needed to dampen the oscillation. The tachogenerator works very effectively for positioning without oscillation. As stated above, though a tachogenerator has a construction similar to that of a DC motor, it is designed to produce an output voltage proportional to its speed. If the output voltage is positive in a clockwise sense, it will be negative in the opposite sense. The effect of the tachogenerator in this positioning system may be explained as follows. When the motor has a speed near the target position, the tachogenerator's voltage is subtracted from the voltage difference between the position command and the potentiometer's output voltage. The voltage applied to the motor is less than before, and this makes the motor decelerate when it approaches the target. Thus, undesirable oscillations do not occur.

As seen in this example, a speed sensor as well as a position sensor is needed in positioning controls. The same is true with the more modern methods explained in the next chapter.

Conclusions

In this chapter, we studied various ways of controlling motor speeds and rotational angles. However, these are regarded as classical methods. In the next chapter, we will see some examples of modern control techniques.

Reference

1. Kenjo, T. (1990), *Power electronics for the microprocessor age*, chap. 3. Oxford University Press, Oxford.

5
Power electronics and modern control methods

In this chapter, we will come across more sophisticated modern methods of motor control. The technology of controlling motors using solid-state devices like transistors, rectifier diodes, etc., is known as power electronics. Several simple configurations using bipolar transistors or a triac have already been looked at in previous chapters. This chapter starts with an explanation of various kinds of solid-state devices for a detailed study.

5.1 Power-electronic devices for motor control

The technology of driving electric motors with solid-state devices is known as 'power electronics'. In power electronics, solid-state devices are used mainly as switches, whereas they are used both as switches and as amplifiers in information electronics. Nowadays, various kinds of solid-state devices are used to control motor torque and speed. Table 5.1 shows the symbols and fundamental characteristics of the main devices. Here, we will study the basic principles and characteristics of various kinds of solid-state devices.

5.1.1 Impurity semiconductor materials and *pn* junction

Most solid-state devices are made of crystalline silicon. Silicon is an element belonging to the fourth column of the periodic table. This means that silicon atoms have four electrons in the valence shell or outermost orbit, as shown in Fig. 5.1(a). Silicon's crystalline structure is similar to diamond's, as illustrated in Fig. 5.1(b). In this structure, each atom shares one of these valence electrons with each of its four neighbours. In such a covalent structure, it is likely that each atom has eight valence electrons. It is known that, when atoms possess eight electrons in their valence shell, they are very stable. Such a crystal structure is often illustrated by the two-dimensional representation shown in Fig. 5.1(c).

Pure or intrinsic silicon is of little value as an electronic device because it has a very low conductivity because of its stable crystalline structure. To be used in a diode or transistor, the crystalline silicon must contain some impurities in small densities.

Table 5.1 Main solid-state devices

Device	Symbol	Characteristics
(1) Rectifier diode		• Simplest device with rectifier characteristics
(2) Bipolar transistor		• Switching is controlled by base current • Low to medium power capability • Majority are *npn* transistors but *pnp* transistors are also used
(3) MOSFET		• Switching is implemented by gate voltage • Power-handling capability is low • High-frequency switching is possible (e.g. as high as 1 MHz) • Simple connection to microprocessor
(4) GTO		• Switched on by trigger current supplied to gate and switched off by drawing current from gate • Suitable for high-power applications
(5) IGBT		• Switching is controlled by gate voltage • Medium power capability • High-frequency switching is possible • Simple connection to microprocessor

If, during the production of single crystal of silicon, some atoms of an element from the fifth column of the periodic table are introduced, these atoms will take up various positions throughout the crystal lattice (see Fig. 5.2(a)). As each impurity atom has four immediate neighbouring silicon atoms, there are four valence electrons from these atoms available for forming covalent bonds, so one of the electrons belonging to the impurity atom is not utilized in a covalent bond. These surplus electrons are loosely bound to the parent atoms and are free to move about at room temperature, which implies that an electric current can flow. The so-called 'doped' silicon crystal shows intermediate characteristics between a conductor and insulator: it is known as a semiconductor. This type of doped semiconductor is called *n*-type, where *n* is derived from the negative charge of the surplus particle.

The other type of doped semiconductor is called a *p*-type material. This is formed by doping with atoms of an element from the third column of the periodic table during production of the crystal (see Fig. 5.2(b)). Since the

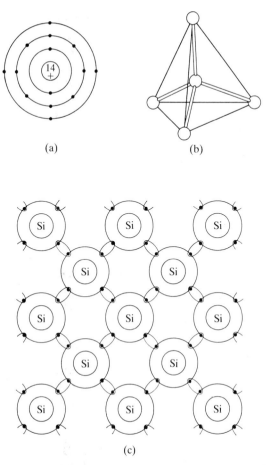

Fig. 5.1 (a) Four of the 14 electrons in a silicon atom are in the valence shell or outermost orbit. (b) The crystal structure of silicon is similar to the very stable diamond structure of carbon: each atom is surrounded by four immediate neighbours. (c) Two-dimensional presentation.

trivalent impurity atom is surrounded by four immediate silicon atoms, one electron is missing from a potential covalent bond. This deficit of an electron behaves as a positive charge carrier, and is known as a hole. Just as with the excess electrons in the n-type impurity material, holes in a p-type material are loosely bound to their parent atoms and are free to move in the material when an electrical field is applied at room temperature.

In most semiconductor devices, both n-type and p-type regions exist in a single crystal. The most frequently used dopant for p-type material is boron (B), while typical dopants for n-type materials are phosphorus (P) and arsenic

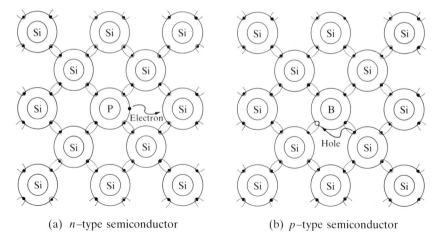

(a) *n*–type semiconductor   (b) *p*–type semiconductor

Fig. 5.2 Silicon crystal with impurity atoms. (a) When pentavalent atoms are inserted, their fifth valence electrons are loosely bound to their parent atoms and can become electrically negative charge carriers. (b) If trivalent atoms are added, the vacancy or deficit of a valent electron (known as a 'hole') can freely move about because of the thermal energy at room temperature and can behave as positive charge carriers.

(As). The transition region from *n*-type to *p*-type is referred to as a *pn* junction. The *pn* junction is very important, and solid-state devices have one or more *pn* junctions.

5.1.2 A *pn* junction as a rectifier

Now, let us study the basic characteristics of a *pn* junction. A diode is a single crystal of silicon that has one side doped with *p*-type impurity atoms and the other side with *n*-type impurities.

It is known that a *pn* junction has the properties of a rectifier, that is, it permits a current to flow in one direction but blocks a current in the opposite direction. A simple but useful explanation of how a *pn* junction exhibits the property of a rectifier is presented in Fig. 5.3. The *p*-side terminal is called the anode, and denoted by A, whereas the *n*-side is the cathode (C). When a positive potential higher than about 0.6 V is applied to the anode with respect to the cathode, the *pn* junction is forward biased and can carry a current. Conversely, when the *pn* junction is reverse biased, that is, a positive voltage lower than 0.6 V or a negative potential is applied to the anode with respect to the cathode, the diode blocks the current.

Thus, forward biasing is equivalent to the ON state and the reverse biasing to the OFF state. Consequently, when a diode is connected as in the circuit

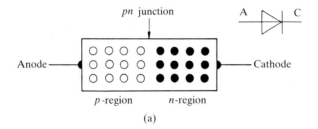

(a)

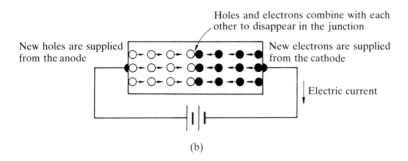

(b)

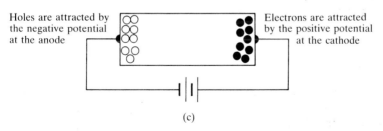

(c)

Fig. 5.3 How a *pn* junction in crystalline silicon works as a rectifier. (a) When no potential is applied to the diode, holes or positively charged particles are free to move in the *p*-region, and electrons or negative particles are free to move in the *n*-region. (b) When forward biased, i.e. when a potential is applied as above, holes and electrons drift toward the *pn* junction owing to the electrical field in each region. In the junction holes and electrons combine with each other to become neutral and disappear. However new holes are supplied from the anode and new electrons are supplied from the cathode. Thus, continuous flows of both sorts of particles are maintained: this is an electric current. (c) When reverse biased, i.e. when a potential is applied as above, holes are attracted by the negative potential at the anode and they are absorbed by it, and electrons are attracted by the positive potential at the cathode and absorbed by it. Thus, all the charge carriers are evacuated from the diode: no current will flow.

Table 5.2   Classification of semiconductor devices based on the $p-n$ structure

Devices	$p-n$ Structures	Symbols	Basic functions
(1) Diode	Anode / p / n / Cathode	A ▽ C	
(2) Bipolar npn transistor	Collector / n / p / n / Base / Emitter	C / B / E	
(3) Bipolar pnp transistor	Collector / p / n / p / Base / Emitter	C / B / E	
(4) GTO	Anode / p / n / p / Gate / n / Cathode	A ▽ C / G	
(5) Triac	Terminal 1 / Gate / n p n / n / p n / Terminal 2	T1 / G / T2	

shown in (1) of Table 5.2, the diode will carry a current in the positive half-cycle of the applied AC voltage and block the current in the negative half-cycle.

5.1.3 Bipolar transistors

A bipolar junction transistor has two *pn* junctions in either a *pnp* or an *npn* construction. No matter which type it belongs to, the central region sandwiched by the two junctions is called the base and denoted by B. One of the two remaining regions is larger than the other, as seen in a 'triple-diffusion planar' transistor illustrated in Fig. 5.4: this region is called the collector, and is denoted by C; the other is the emitter (E).

Figure 5.5 illustrates an *npn* transistor connected in a 'common emitter' scheme. In this type of connection, the base is used for the input terminal and the collector for the output terminal, while the emitter is common both to the input and output stages. A relatively low forward potential E_1 plus an alternating signal potential is applied between the base and emitter. The battery potential E_2 on the other side is higher than E_1. Hence, a reverse potential is applied across the *pn* junction between the collector and base. Since the *pn* junction between B and E is forward biased, free electrons enter the base region from the emitter.

It is very important that the base region is made sufficiently thin so that most electrons coming from the emitter can penetrate the base and enter the

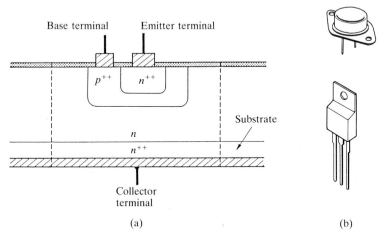

(a)   (b)

Fig. 5.4 (a) Cross-sectional structure of one unit of a 'triple-diffusion planar' transistor fabricated on a highly doped substrate denoted by n^{++}. Collector area is much larger than the emitter. Hundreds or thousands of such units are fabricated on a substrate and they are connected in parallel to be able to handle high currents; (b) examples of transistors.

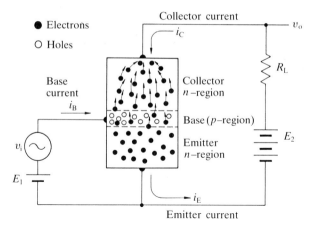

Fig. 5.5 Movement of electrons and holes in the common-emitter connection when a positive voltage is applied to the base with respect to the emitter. Most electrons injected from the emitter to base region travel by diffusion towards the collector region to produce the collector current. Some electrons recombine with holes in the base region. To supply the holes lost, a current flows into the base.

collector region. In this region, the electrons are accelerated towards the collector terminal by the reverse potential of E_2. When the signal potential v_i is higher, more electrons will travel from the emitter to the collector region producing more current. On the other hand, when v_i is made lower than 0.6 V or negative to reverse bias the base–emitter junction, no electrons will travel towards the base or the collector, and no collector current will flow.

As stated above, the base region is so thin that most electrons injected from the emitter to the base enter the collector region, having no opportunity of combining with a hole. However, the probability of recombination between a hole and an electron in the base is not absolutely zero. There are some electrons and holes which are lost because of recombination. To supply the holes to the base region, a current flows from the input power supply (E_1 and v_i) towards the base: this is the base current.

5.1.4 Linear and ON/OFF operations

When the collector current varies with time, the potential across the load resistor R_L wlll vary too. Figure 5.6 illustrates two cases of relations between v_i and v_o: one is for a sinusoidally or continuously varying input signal and the other is a square-wave input signal.

When a transistor is used as an amplifier in the continuous mode to drive a motor, the simple common emitter connection is not employed because, with this connection, parameters like the current amplification factor differ

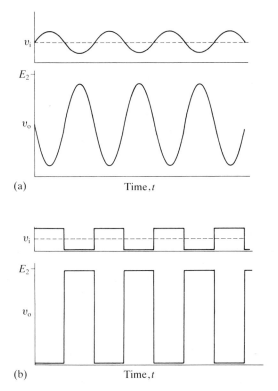

(a)   Time, t

(b)   Time, t

Fig. 5.6 Relationship between the input signal v_i and output waveforms v_0: (a) sinusoidal input signal; (b) square-wave signal.

from transistor to transistor and are strongly affected by temperature. The servoamplifier circuits shown in Figs 4.14 (p. 111) and 4.15 (p. 112) are typical examples using bipolar transistors as amplifiers. In technical terms, we say that transistors are operated in the linear region.

The latter method is the switching mode. Look again at Fig. 3.45 (p. 90) that was used to explain the similarity between a transistor switch and a mechanical switch in Section 3.7. When a voltage is not applied to the base with respect to the emitter, hence no current is supplied to the base, the transistor acts as an open switch, while if a voltage higher than 0.6 V is applied to the base or a sufficient base current is provided it functions as a closed switch.

In current motor-control techniques, bipolar transistors are used most often as solid-state switches. The bipolar transistor features a high current density per unit area of the semiconductor material.

The difference between *npn* and *pnp* transistors is in the bias polarities, and a comparison of these is given in Table 5.2.

5.1.5 MOSFETs

The metal-oxide field-effect transistor, or MOSFET, used as a switching element in power-electronic circuits, is typically constructed as shown in Fig. 5.7(a). The gate is the control terminal, which corresponds to the base of a bipolar transistor. One of the features of the MOSFET is that the gate is

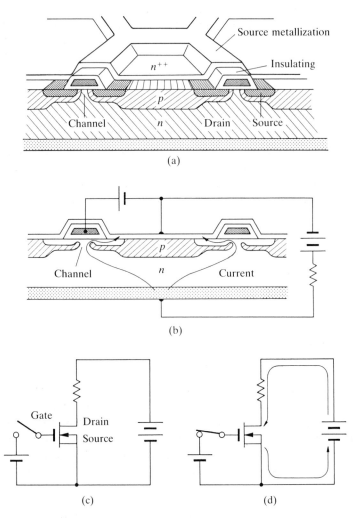

Fig. 5.7 Typical construction of MOSFETs and their principle of operation. (a) Cross-section of a MOSFET when no potential is applied. Channels are dosed by the *p*-region. (b) When a positive potential is applied to the gate with respect to the source, the channels are opened. (c) When no potential is applied to the gate, the MOSFET does not carry a current. (d) When a positive potential is applied to the gate with respect to the source, a current can flow from drain to source.

isolated, by an insulator film, from the source and drain, which correspond to the emitter and collector in a bipolar transistor, and hence no appreciable current flows in the gate terminal. The drain electrode is in contact with the n-type substrate, and the source metallization is in contact with both an n-type and a p-type region.

The construction which we are discussing here is the 'n-channel' MOSFET. How channels behave in this construction can be explained using Fig. 5.7(b), as follows. When no potential is applied to the gate with respect to the source, the n-regions of the source and the drain are separated by the p-region beneath the gate. In this state, no electrons can travel from source to drain even if a positive potential is applied between the source and the drain. When a positive potential is applied to the gate, however, the holes in the p-region beneath the gate are depleted by the electrical field. This results in expansion of the n-region and shrinkage of the p-region to form an n-channel, through which current can flow from drain to source.

Thus, while in a bipolar transistor there are two pn junctions in the main current path, there are no pn junctions in the current channel of a MOSFET. So the MOSFET is a monopolar device. MOSFETs are superior to bipolar transistors of comparable size in switching-speed performance.

5.1.6 Insulated-gate bipolar transistor (IGBT)

A bipolar transistor features high current density, while a MOSFET has the advantage of being operated by a high-response voltage and being more compatible for interfacing with a microprocessor. The insulated-gate bipolar transistor or IGBT is a monolithic device created to combine the merits of these two different types of devices.

5.1.7 Gate turn-off thyristor (GTO)

A GTO has three pn junctions and three terminals: an anode (A), a cathode (C), and a gate (G), as shown in Table 5.2. The gate is the control terminal, but its use is different from that of the MOSFET. When a trigger current or a current of very short duration is applied to the gate, the GTO fires and a large current flows from A to C. To turn off the GTO, a negative current is applied to the gate. GTOs feature high current and very high voltage capabilities.

5.1.8 Triac

A triac is a bidirectional device used in an AC circuit. The junction structure, symbol, and basic operation are illustrated in Table 5.2, and also in Fig. 4.4 in the previous chapter. A triac can be turned on either by supplying or drawing a current from the gate when a positive or negative voltage is applied across terminals 1 and 2. The triac switches off automatically when the current falls to zero. By varying the timing of the trigger signals to the gate, we are able to control the AC current at the load. It has already been explained that the speed of a single-phase AC motor can be regulated by this technique.

5.1.9 Trigger diode

A trigger diode is a tiny device designed to generate trigger current to switch on a triac, as used in the conduction-angle control in Fig. 4.3 (p. 101) in the previous chapter. It has a simple symmetrical *pnp* structure with two electrodes, as shown in Fig. 5.8(a). When the voltage between the two electrodes is increased, little current flows until a certain critical level is reached. When the voltage exceeds this critical value, a so-called breakdown occurs inside the *pn* junction and a large current flows. This current is suitable for use as the triggering current for a triac. The characteristics are ideally symmetrical, as illustrated in Fig. 5.8(b).

5.1.10 Use of microprocessor and digital devices

Figure 5.9(a) shows a solid-state power circuit designed for multipurpose use suited to experiments using the stator and rotors shown in Chapter 3 (see Figs 3.3 and 3.11). As shown in the circuit diagram in Fig. 5.9(b), this circuit has four pairs of arms, and each arm consists of an *npn* bipolar transistor for

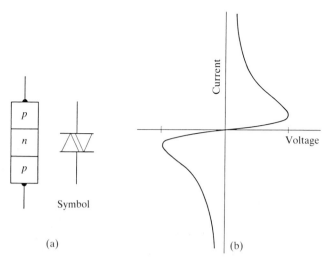

Fig. 5.8 (a) Junction structure of trigger diode; (b) current-versus-voltage characteristics.

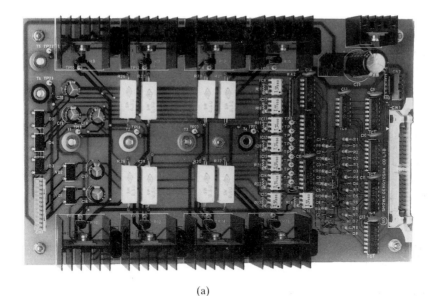

(a)

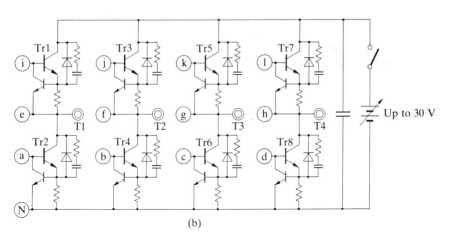

(b)

Fig. 5.9   A universal power circuit using bipolar transistors: (a) appearance; (b) the main circuit.

switching use and a smaller transistor to work as a current limiter in association with the resistor connected to the main transistor's emitter. The resistor
senses the electric current, and, if the current exceeds a certain value, the
smaller transistor closes and the collector-to-base potential at the smaller
transistor becomes negative. Note that this voltage is applied across the base
and emitter of the main transistor. As explained using Fig. 5.5, when a negative

Fig. 5.10 Microprocessor board for generating switching signals to be supplied
to the power circuit of Fig. 5.9.

voltage is applied to the base with respect to the emitter, the main transistor
opens. If this current limiter is not included, the transistor will burn up when
a current over the rated level flows in it.

Switching signals are generated from the control board shown in Fig. 5.10,
and transmitted to the power circuit via optical devices and integrated logic
circuits. The control board consists of a microprocessor 8085A and some
associated integrated circuit chips. Figure 5.11 is the circuit diagram presented
for those who have a special interest in microprocessor circuit designs. The
software for driving motors is incorporated in the memory chip, type number
2764.

Several typical connections between the power circuit and motors are
illustrated in Fig. 5.12.

5.2 Pulse-width control for saving energy

In the examples of speed control and position control of a DC motor given
in the previous chapters, the voltage applied to the motor was varied continuously, and the transistors were operated in the linear mode. The drawback
with this method is a big energy loss in transistors: a big heat sink is needed.

A modern method of controlling voltage is by so-called pulse-width modulation, or PWM for short. The principle of PWM is very similar to that of
the mechanical governor discussed in Section 4.5.

'Governor' has various meanings, but here it means a voltage-control
method, turning the circuit ON and OFF at a high frequency, as shown in

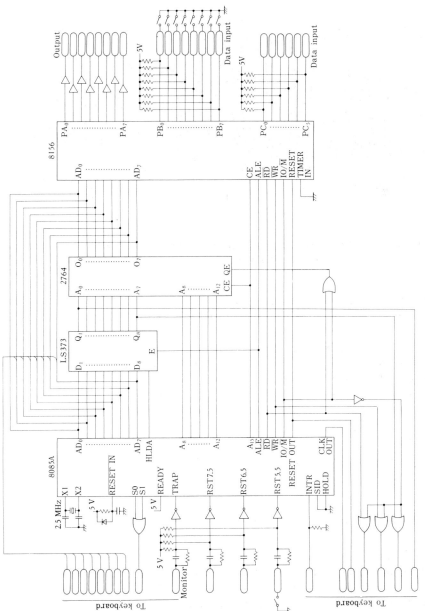

Fig. 5.11 Circuit diagram of the microprocessor board.

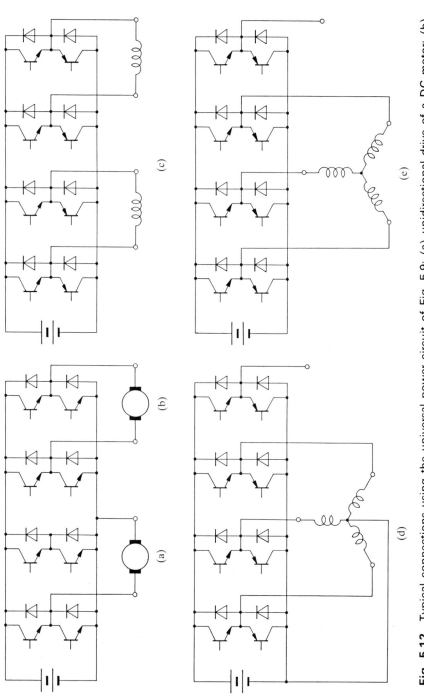

Fig. 5.12 Typical connections using the universal power circuit of Fig. 5.9: (a) unidirectional drive of a DC motor; (b) bidirectional drive of DC motor; (c) bipolar operation of a two-phase stepping motor; (d) unipolar operation of a three-phase stepping motor; (e) inverter operation of a three-phase AC motor, and brushless DC motor with position sensor feedback.

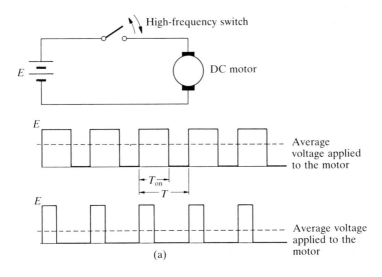

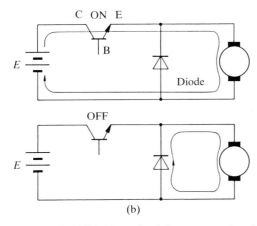

Fig. 5.13 Principle of PWM drive of a DC motor, and a circuit using a bipolar transistor: (a) using a mechanical governor; (b) using a bipolar transistor (a rectifier diode is placed in parallel with the motor to provide a path for the current which results from the inductance of the motor when the transistor is OFF).

Fig. 5.13. Here, let the time for the switch to close be T_{on} and the time for it to open be T_{off}. If the power-supply voltage is 10 V, and the ratio of T_{on} and T_{off} is 6 : 4, the average voltage applied to the motor terminals is 6 V. When the ratio is changed to 3 : 7, an average voltage of 3 V will be applied to the motor.

It is difficult to control mechanical contact at a high frequency, and, even if it is possible, it is not useful because the switch will soon wear. We therefore use a semiconductor switching element like a transistor or MOSFET, as shown in Fig. 5.13(b). Here, a diode and a coil are additionally used in this configuration so that the current flows continuously to the motor. In this circuit, the switching-signal train is applied to the transistor's base (B) terminal. When a positive voltage is applied to B with respect to the emitter (E) terminal, the collector (C) and the emitter (E) are closed, so that the battery voltage is applied at the motor. Conversely, if the base voltage with respect to the emitter is zero, the C and E terminals work as an open switch, and no voltage is applied to the motor. The average voltage $\langle V \rangle$ across the motor terminal is

$$\langle V \rangle = ET_{on}/(T_{on} + T_{off}) = E \frac{T_{on}}{T_{on} + T_{off}}. \tag{5.1}$$

In practice, the switching is done 1000 to 20 000 times a second. When the switching frequency is low, the motor will cog and audible noise will be generated from the motor and circuit components. However, if the frequency is higher than 16 kHz, there are no problems.

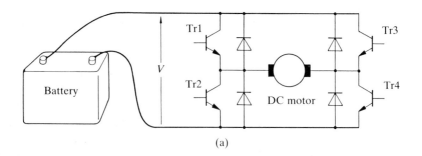

(a)

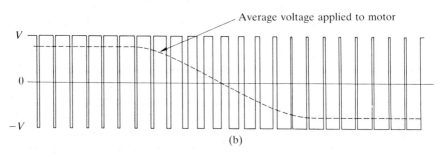

(b)

Fig. 5.14 (a) H-bridge circuit for reversible (bidirectional) operation of a DC motor; (b) the voltage waveform applied to the motor.

When one uses the universal power circuit shown in Fig. 5.9 for this experiment, the circuit will be like that shown in Fig. 5.12(a). A feature of the PWM technique is that the electric power consumed in the transistor is much lower than that which occurs in linear operation.

A circuit called an H-bridge, which consists of four transistors arranged as shown in Fig. 5.14(a), can be used as a reversible drive of a DC motor in the PWM mode. These transistors can be replaced by MOSFETs or IGBTs. Several different switching sequences are possible for the H-bridge, but the simplest is turning Tr1 and Tr2 ON/OFF at the same time, and Tr3 and Tr4 in the opposite state. The waveform across the motor terminals is as shown in Fig. 5.14(b), where it is seen that the average voltage is continuously varying in accordance with pulse width.

5.3 Pulse-width-modulated three-phase inverter

In the inverter described in Section 4.2, the waveform across any two phases of a three-phase motor was rectangular and the voltage was determined by the DC battery voltage only. However, if we apply the PWM technique to a three-phase AC motor using the connection shown in Fig. 5.12(e), the voltage waveform can be as close to a sinusoidal wave as possible, and its effective voltage can also be controlled.

Figure 5.15(a) shows an example. Here, what is most important is the line-to-line waveform or the voltage appearing across any two of the three phases. It is the waveform shown by the notation V_{a-b}. It is interesting that the voltage waveform of each phase with respect to the ground terminal is not a sinusoidal wave. It should be noted that the line-to-line waveform contains a sine wave, which means that the average voltage calculated according to equation (5.1) will yield sine plots. Figure 5.15(b) shows the current waveform at an eddy-current motor driven in this manner. If the number of pulses is increased, the associated ripple component will be reduced and the current will be very close to a sine curve.

The so-called variable-voltage variable-frequency (VVVF) drive of an induction motor uses the PWM technique. When the frequency is varied in proportion to voltage, the torque-versus-speed characteristics of an induction motor are as shown in Fig. 5.16. Compare these with the curves in Fig. 3.22 (p. 70), which are the characteristics obtained by varying rotor resistances. In the VVVF drive, both frequency and voltage are set low at the start and they are increased as the motor accelerates. Since the high-efficiency operating points are always utilized in variable-speed drives, with this method the overall energy utility is excellent.

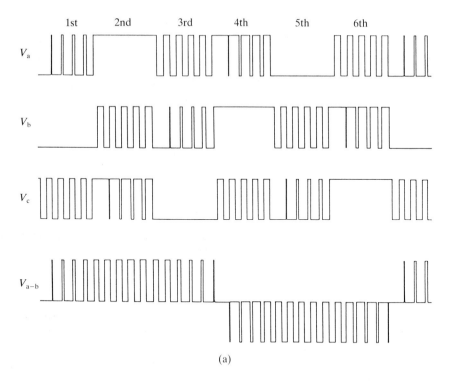

(a)

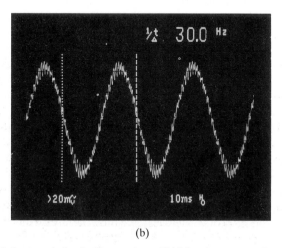

(b)

Fig. 5.15   (a) An example of a three-phase PWM voltage waveform; (b) current
waveform when a solid-steel eddy-current motor is running.

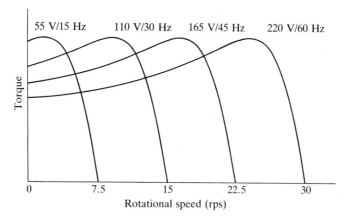

Fig. 5.16 Torque-versus-speed characteristics of an induction motor driven by a VVVF inverter.

5.4 Servomotors

Servomotors are a kind of control motor, or motors used as the motive source in a servosystem for the positioning and speed control of robot arms or the mechanism in numerically controlled machines. In terms of machine structure, there are both DC servomotors and AC servomotors.

Servomotors in a robot are analogous to muscles in the human body. Although a human being has more than 500 muscles, a robot has far fewer motors. However, it is not very easy to control several servomotors in a numerically controlled machine.

While a general power-use motor is designed to turn at basically one speed, servomotors are designed to carry out operations following a wide range of speed instructions. The word 'servo' comes from the Latin *servus* meaning 'slave', and a servomotor can be thought of as a motor that works by faithfully following its master's 'orders'. Here, the 'orders' are position or speed instructions. Sometimes, the order can be a combination of position and speed, for example, if the position command is being varied continuously in a position-control scheme, the motor is subjected to speed control. Accordingly, a servomotor must have the following characteristics:

(1) to turn stably over a wide range of speeds; and

(2) to change speed swiftly (in other words, to generate a high torque from a small size).

Motors used in a feedback loop for constant-speed operation (e.g. the spindle-drive motor in a disk drive) are usually not regarded as servomotors, because these motors do not need to respond quickly.

5.5 Servosystems using microprocessors

In the classical positioning technique which uses a potentiometer, described in Section 4.10 in the previous chapter, high positioning accuracy is not always expected because the position information is measured in analog quantities. For instance, if any changes occur in the biasing voltage to the potentiometer, the output voltage varies in accordance with that change, and this appears to be a position deviation. For this reason, a digital position sensor like a photoencoder is used for applications that need higher positioning accuracy.

Another important position sensor is called a resolver, which is very similar to an AC motor in machine construction but uses the transformer principle. The position information is originally obtained in voltages from two windings, and then converted to digital values.

As discussed in Section 4.10, speed information is needed in positioning control, and this is true with advanced methods. There are basically two ways of obtaining the speed information: one is from the position signal and the other is to use a tachogenerator.

Figure 5.17 shows the principle of an XY table, which is a typical position-

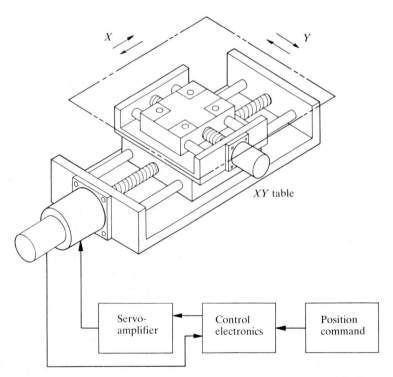

Fig. 5.17 Principle of an *XY* table, a typical position-control device.

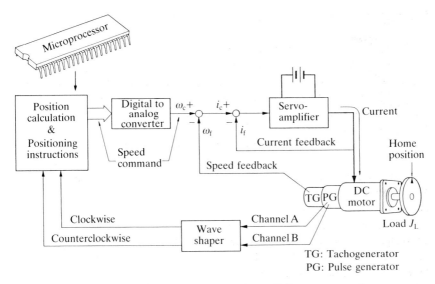

Fig. 5.18 Positioning servomechanism using a DC motor, a pulse generator (PG), a tachogenerator (TG), and microprocessor.

ing device used in mechanical workshops. Movement in the X-axis direction is controlled by one motor, and in the Y-axis direction by another. We consider the control of the X-axis driven by a DC servomotor. When a positioning instruction is given to the control electronics, an appropriate current command will be computed from the present position information feedback, and this is sent to the servoamplifier which drives the motor.

Figure 5.18 is a more detailed block diagram that explains the function of this control technique. The DC servomotor is mounted with an encoder and a tachogenerator. A microprocessor, which is a compact digital computer, is used to compute the motor's present position (or rotational angle) by counting the pulses coming from the encoder, and also to generate the speed command as a function of the distance from the present position to the target position. The digital-to-analog converter is necessary to convert the speed command to a voltage to be compared with the speed feedback quantity coming from the tachogenerator. The difference between the speed command and the speed feedback quantities is used as the current command, and it is compared with the current information voltage. The servoamplifier works to drive the motor at the speed that makes the current equal to the current command thus yielded.

The mathematics behind this system is as follows. The distance travelled or the position from a reference position is the integration of the speed with respect to time. Therefore, to pursue a position control, both position information and speed information are needed.

5.6 Fully digital control

In the previous example, the speed information is generated as an analog value, since it is detected by a tachogenerator. If the speed is calculated from the pulse intervals using a digital logic circuit and a crystal oscillator, the control system does not need a digital-to-analog converter. Such a fully digital configuration tends to be employed more than the hybrid configuration.

In a control system using analog variables for speed or position, a lot of resistors used in the control circuitry must be adjusted in the final manufacturing stages to attain smooth vibration-free motion. After installation, there is always a possibility that readjustment is required due to changes in resistance owing to ambient temperature or ageing. If all such resistors were eliminated, the manufacture and service cost would be reduced, and the system reliability upgraded.

In a fully digital system, one or more microprocessors are used. This means that software and hardware advantages are both utilized. In a hardware-oriented system, to change a parameter means to replace a capacitor or a resistor with another using a soldering iron. This is laborious and can degrade system reliability. In a software-oriented system, the corresponding procedure can be done by rewriting a program statement stored in a memory chip. It is also unrealistic to eliminate hardware logic or arithmetic, because hardware computation is very quick and some logic is much simpler using hardware.

The other advantage of using software is that diagnostics is available. For example, when a system failure has occurred, we can test the system to find out the defective part by a previously provided program.

5.7 Necessity for brushless servomotors

The control system using a DC servomotor discussed in Section 4.10 is not complicated. However, note again that a DC motor has brushes and a commutator and that these components can cause trouble. As previously explained in detail, during operation the brushes and commutator segments are sliding at high speed. Since the main component of the brush is carbon, it can wear out after a long running time.

The wearing time depends on humidity and some other factors to do with the atmosphere. When the rotating speed is high, wear proceeds quickly, and hence the maximum speed is limited. Therefore, before the brushes have worn out, they must be replaced with new ones. The number of motors used in the old days was small, but there were technicians who had good experience in the maintenance of brushes and commutators. Recently, the number of servomotors in a machine has increased, but the number of qualified technicians has inversely decreased. Moreover, some machines have a construction in which replacing the brushes is impractical. Furthermore, maintenance cost is high, because the machine must stop operation.

For these reasons, more brushless DC motors are used nowadays. As discussed in Section 3.7, normal brushless DC motors have three-phase windings and use a three-phase inverter for operation. For a type required to move with a high-response capability, a thin permanent-magnet rotor as shown in Fig. 5.19 is used. When a brushless DC motor is used as a servomotor, the inverter is used as a servoamplifier. Of course, when the inverter is used as a servoamplifier, it is used to control each current in the three phases. Naturally, the power circuit is more complicated than that for a conventional DC motor.

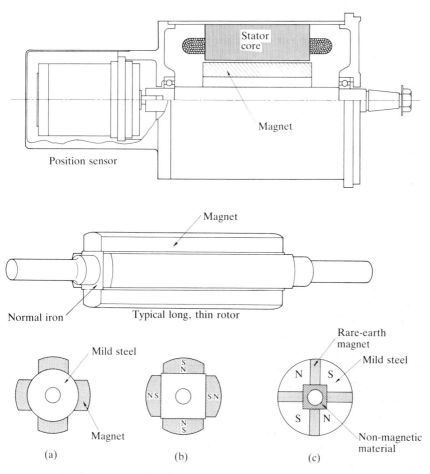

Fig. 5.19   An example of the construction of a brushless DC motor.

Powerful permanent magnets tend to be used in the rotors of brushless DC motors to generate a strong torque from a small volume.

5.8 Vector control of induction motors

Squirrel-cage induction motors are also used as servomotors. Before the computer came to be used to control the induction motor, it had been regarded as a slow-response motor. However, if the current in each phase is carefully controlled by speed commands from an inverter, the induction motor can change its speed very quickly like a DC servomotor. With this method, a lot of computation is done during operation, based on feedback information about motor speed, rotational angle, winding currents, and machine parameters. This control concept is known as 'vector control'. In mathematical terms, a vector is a quantity that has a direction and a magnitude. In vector control of an induction motor, the three-phase voltages and currents are dealt with as vectorial quantities.

Vector control is suited to large machines such as those for turning the rollers in an iron mill, while brushless DC motors with a permanent-magnet rotor are more suited to medium- and low-power applications.

5.9 Smartpower integrated circuits for motor drive/control

As we have seen above, most stepping motors are controlled without feedback, but brushless DC motors are normally operated with feedback controls. The control signals driving the solid-state devices in a motor drive system are generated by an electronic circuit consisting of digital integrated circuits (ICs) and/or a microprocessor. To drive a medium-size or large motor, a power circuit consisting of transistors or other types of solid-state devices and a control circuit are separate, and electrical signals are transmitted between the two sections via a cable or a printed-circuit board.

However, for small motor applications, the technology for manufacturing both power circuitry and the control or smart portion on a single chip (substrate) has recently been developed. This type of integrated circuit is called a 'smartpower' IC.

As shown in Fig. 5.20, smartpower modules can also contain sensors to measure temperature, currents, and voltages. This information is used for diagnostics and protection of the module itself and other components.

Smartpower ICs are used in large quantities for the speed-controlled micromotors in cooling fans, VCRs, and office machines such as floppy disk drives. Many stepping motors used in disk drives and printers are also driven by smartpower ICs. One large application area of smartpower ICs is the control of DC micromotors in an automobile. It is expected that highly intelligent, higher-power smart ICs will be manufactured in the future.

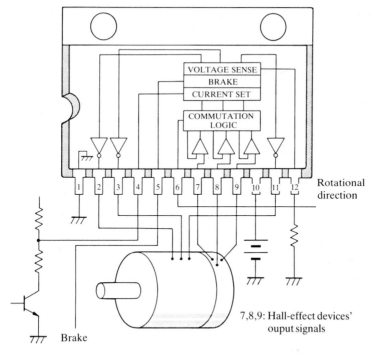

VOLTAGE SENSE
BRAKE
CURRENT SET
COMMUTATION LOGIC

Rotational direction

7,8,9: Hall-effect devices' ouput signals

Brake

Fig. 5.20   Smartpower IC for motor operation.

5.10 Emerging new problems

The PWM technique is advantageous because it generates only low power losses in solid-state devices, and the equipment's size can be made small. However, high-frequency switching always causes electromagnetic noise. When such a driving circuit operates near a radio receiver, the sound is often contaminated by interference due to the high-frequency pulse modulation. Electric noise also propagates along power network lines and causes an increase of power loss in the transmission system. New studies aimed at solving these problems have recently become very important.

Conclusions

We have seen what power electronics for motor control is like. Pulse-width modulation using solid-state devices is an important technology, and, based on it, highly intelligent controls are being developed. However, it was also pointed out that the PWM method generates electromagnetic interference noise and produces an undesirable effect in the power network. How to get rid of these adverse effects is a new up-and-coming problem for engineers.

6
Small motors and the info-society

In this chapter, we will look at what is going on with conventional motors and the latest motors to deepen our insight into what motor science and technology means in the high-tech info-society.

6.1 Single-phase induction motors in automated factory machines

A passage in the preface of Professor Tim Miller's monograph[1] says:

> If nothing else, a study of brushless motor drives will lead to a further appreciation of the extraordinary properties of conventional motors, the DC commutator motor, and the AC induction motor.

Indeed, squirrel-cage induction motors are predominantly used in factories, houses, and hospitals. Most power motors are orthodox three-phase machines, and many of them are driven via an inverter with electronic controls.

However, many more small single-phase motors, which have a rugged construction, are used in automated equipment with a variety of designs. Some types are suited to continuous running at a regulated speed with simple electronics, some are for reversible operations, and some are for a high-torque output. Figure 6.1 shows motors with a gear-head and their speed controllers. Small, slim fans incorporated in electronic equipment are also driven by induction motors.

Oriental Motor, a Japanese manufacturer, is an extraordinary manufacturer of some thousand different standard types of single-phase induction motors. They never have such a large variety in stock, but when an order for any number of any type is received, they can be manufactured and delivered within 24 hours to any corner of Japan. The world's first single-phase induction motor was built in the USA by Nikola Tesla, a Yugoslavian–American scientist, who died poor in 1943 in New York City. Oriental Motor and the Japanese motor industry must extend deepest thanks to this unhappy inventor.

6.2 Low-cost mass-produced small DC motors

Small DC motors are manufactured in large numbers and also in a wide range of designs. Even the most popular toy motor, whose typical construction is

Fig. 6.1 Single-phase induction motors (Courtesy Oriental Motor Co., Ltd.).

shown in Fig. 6.2, is not as simple as a stepping motor or an eddy-current motor. However, the DC motor is the lowest-cost choice for a low-voltage DC-powered application, because a stepping motor needs an electronic circuit to turn it and, for driving a squirrel-cage induction motor from a DC power supply, an expensive inverter is absolutely necessary. Their energy-conversion efficiency is much lower in small motors.

In particular, DC motors with permanent magnets produce a high torque with low heat loss. Moreover, it is a great advantage that the motor terminals can simply be connected to a battery in order to run it. Consequently, DC motors are used exclusively in toys and automobiles; dry cells are used in toys, but lead batteries are used in automobiles. Plain DC motors are also used in large numbers in VCRs, photocopiers, and cassette tape recorders.

The Japanese company Mabuchi manufactures two million motors a day at their factories overseas! If they were lined up, the total length would reach 50 km, assuming an average motor diameter of 25 mm. However, production speed is not very fast if one notes that a distance of 42.2 km can be run by a good marathon runner in a little over two hours. Two million motors are manufactured by many lines in two shifts (about 16 hours). As demand increases day by day, the mass-production speed is getting close to that of the marathon runner.

As was often pointed out in the previous chapters, the only drawback of the DC motor is the short lifespan due to the sliding contact between brushes

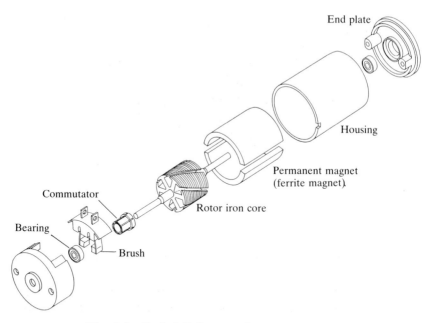

Fig. 6.2   Exploded diagram of a small DC motor.

and commutator. However, by placing an element called a 'varistor' between two adjacent commutator segments, sparks are reduced and the lifespan can be doubled.

Apart from that, there are a lot of application areas that only need a lifespan of 1000 hours. (Most DC motors can run for a longer period.) For example, if a car is driven for two hours a day, and if the probability of rain is 10 per cent, the total driving time in rain is only 73 hours a year. That means that the windscreen-wiper motor will last for 13 years.

Today, it is rare to hear of academics whose subject of study is small DC motors. However, if someone is successful in creating a new simple DC motor mechanism that is free from spark and wear, it would surely be a great invention. Nobody has yet succeeded in doing this.

6.3 Coreless, moving-coil, and printed-circuit motors

Coreless and moving-coil motors are high-performance DC motors. The rotor carries only armature coils and a commutator; it has no iron core except for the shaft. The coreless and moving-coil motors have a very similar rotor construction. However, usually, the type shown in Fig. 4.7 is considered a typical moving-coil motor. In this type of motor, strong permanent magnets

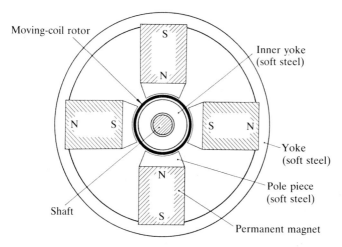

Fig. 6.3   Cross-section of a moving-coil motor.

are placed outside the rotor to concentrate the magnetic field in the air gap. In the hollow space of the rotor winding is a stationary iron core as shown in Fig. 6.3. Since the rotor is light, the movement is very quick, and hence the moving-coil motor is suitable as a servomotor. As a very strong torque is applied to the rotor wire, the windings are reinforced with a strong resin and glass fibre.

The coreless motor is a micromotor with the construction shown in Fig. 6.4. The field magnet is placed in a hollow space inside the rotor. The housing works as a magnetic-flux path. Since the magnetic flux is not very dense because of the small size of magnet, a quick response like that of a moving-coil motor cannot be expected. Coreless DC motors are used in hand machines like cassette tape recorders and also in high-precision instruments.

Why is a coreless motor superior to a conventional DC motor with a slotted core? First of all, it is suited to dry-cell battery operation because of a very low voltage drop across the precious-metal commutator and brushes. The metal commutator and brush cannot be used in a motor with a slotted core because of strong sparking due to the electric parameter associated with flux ripples in the air gap. Slotted-core motors use carbon (graphite) brushes to suppress the spark; the combination of graphite and copper is congenial and has a long lifespan.

If there are no slots, and only moving-coil windings are in motion in an air gap, there is less spark generation, and the use of metals for the brushes and commutator becomes feasible. However, metals such as aluminium and iron are chemically so weak that erosion can occur during operation. Precious metals such as platinum, silver, and vanadium are used. Another merit of the coreless motor is that the torque ripple is very low.

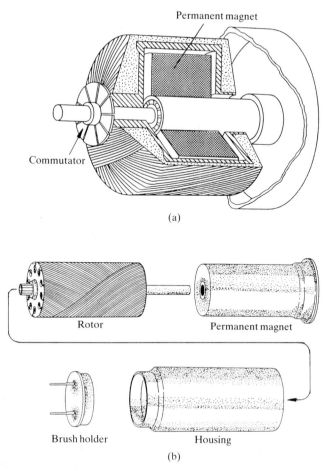

Fig. 6.4 Coreless micromotors: (a) ball winding type; (b) honey comb winding type.

Another version of the coreless motor is the printed-circuit motor, which has a disc-shaped rotor, as shown in Fig. 6.5. The rotor winding was made in the same chemical process as a printed-circuit board when it was invented in 1966 by the French inventor J. Henry-Baudot. Later, it came to be fabricated by a stamping and welding process to minimize manufacturing time. The name 'printed-circuit motor' has remained. (It is also known as the 'pancake motor'.)

6.4 Some small motors and the largest motor

This book is concerned mainly with small motors, and we have claimed above that this high-tech info-society is run by a bunch of small motors. But what

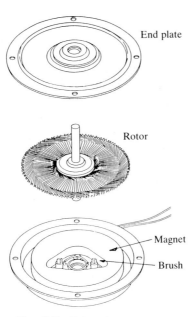

Fig. 6.5 Printed-circuit motor.

is the definition of a 'small motor'? Those who are working on the tiny motors used in watches claim that the motors used in a robot are large. In contrast, engineers dealing with motors driving the rollers in an iron mill feel that a 5 horse-power induction motor is a small motor.

Some feel that a 'fractional horse-power' motor is a synonym for a 'small' motor. By the way, what is a horse power? This unit was used very often, but it is now an old term, because it is excluded from the SI system of units. The equivalent of 1 horse-power in SI units is 750 watts (W).

If a motor continuously consumes 4.5 A of current when it is supplied from a 220 V single-phase net, it consumes 4.5 × 220 = 990 W of electric power. And, if 24 per cent is consumed as heat loss inside the motor and the remaining 76 per cent is converted to the mechanical output, the output power is 990 × 0.76 = 752 W, a little higher than 1 horse-power. The motor used in a vacuum cleaner has about a 1 horse-power capability. Do not associate this unit with the force that a racehorse can exert.

Another definition of a small motor is any motor that can be held in the palm of one's hand, which is comparable to the idea of a fractional horse-power motor. However, the control technology of servomotors ranging from 30 to 3000 W is thought to fall into the category of small motors.

What is the smallest motor ordinarily available? It is the stepping motor turning the arms of a watch, described in Chapter 3. On the other hand, how large is the largest motor in the world? A large rolling motor has 20 million watts of output power. However, the largest is a type of hydroelectric generator of some 5 billion watts. When there is a great demand for electricity, the hydroelectric generator works as a generator powered by water falling from the upper dam. However, when the demand is low, at night, the machine is operated as an electrical motor and pumps the water up from the reservoir below to the dam above, using power supplied from distant thermal power stations running without stopping for a year or so to maintain the high machine efficiency.

6.5 Stepping motors with fine teeth

Of the various kinds of motors, the stepping motor is particularly suited to digital-control electronics. It is widely used for positioning the magnetic reading/writing head in floppy and hard disk drives. As explained in the preceding chapters, there are three major types: the variable-reluctance, hybrid, and permanent-magnet claw-tooth stepping motors. The hybrid type features a small step angle or a high rotational resolution. In order to increase the data volume in a magnetic memory disk, it is necessary to narrow the track width and increase the positioning accuracy. To meet this requirement with a stepping motor, the number of teeth on the rotor and stator cores must be increased and precision machining must be done. The accuracy depends on the accuracy of the die used to stamp out the cores.

Figure 6.6 shows the stator of a 400-step motor used in a hard disk drive. There are 16 poles and on each of them five small teeth are cut; the core's inner

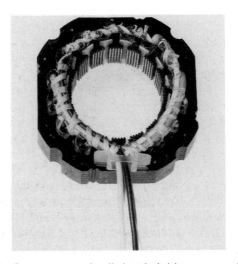

Fig. 6.6 Stator core and coils in a hybrid-type stepping motor.

diameter is 20 mm. The step angle of this machine is 0.9°, and positioning accuracy must be within several percentage points. This means that allowable error is several micrometres. These dies are built by talented, well-trained technicians. Recruiting young gifted technicians and giving them good training is a serious problem for tomorrow's high-quality production, as young technical men prefer to work with computers nowadays.

The low-cost type of stepping motor is the claw-tooth, and its miniaturization is still proceeding. A motor with a diameter of 12 mm has recently become available.

The stepping motor's early history was in Britain: basic inventions and fundamental studies were made by British engineers and scholars. Next, the Americans played an important role in the advancement of applications connected with numerical controls. In the scientific scanning equipment in American spaceships, stepping motors are used as the actuators. They are remotely controlled from ten million to several hundred million kilometres away at the station on Earth. Recently, Japanese engineers have displayed their expertise in the quality control of mass production, miniaturization, and high accuracy of these motors.

6.6 Brushless DC motors in information equipment

Various applications in computers and other information-technology products require constant-speed drives. In these areas, brushless motors are mostly used. Examples are: record-player turntables; tape-recorder and VCR capstans; the rotary heads of VCRs and DATs (digital audio tape recorders); CD players; floppy and hard disks; the polygonal mirror of a laser printer; and cooling fans.

A characteristic feature of these motors is that the construction is quite different from those of conventional motors. For example, a motor is built up on a printed-circuit board, in a rotary head, or in a hub supporting hard disks. In the example shown in Fig. 6.7, six coils, the driving circuit, Hall-effect devices, and more than 30 parts are built into the board.

Figure 6.8 shows a cooling fan for electrical devices. It is the outer-rotor type, and the fan itself acts as an extended rotor. Not only the stator core and coils but also the driving circuit is built inside the cylindrical magnet of the rotor.

Thus, a motor is not always manufactured as one unit. When a variety of components have been assembled on a board, a motor section is incorporated in it. The boundary between the motor and other functional parts is ill-defined. The coils are merely major components and some coils for micromotors are manufactured by chemical processes. Figure 6.9 shows the cross-sectional diagram of a 4 mm-thick motor designed to drive the reels and capstan in a

Fig. 6.7 Brushless DC motor assembled on a printed-circuit board (Courtesy Shinano Kenshi Co., Ltd.).

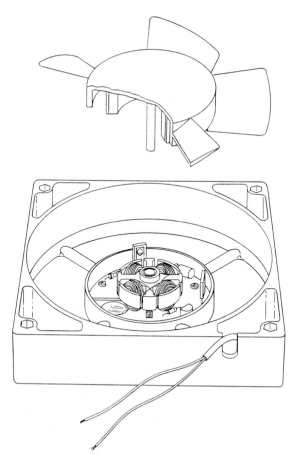

Fig. 6.8 Brushless DC motor assembled in a cooling fan.

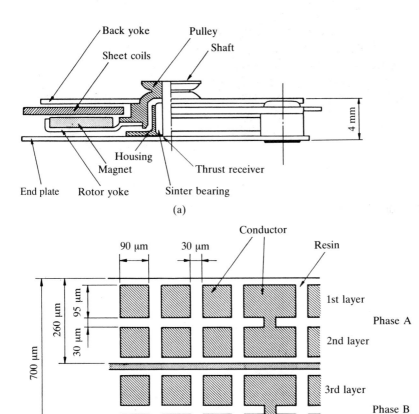

Fig. 6.9 Flat brushless motor using sheet coils: (a) cross-section of motor; (b) cross-section of sheet coils.

small cassette tape recorder. The winding structure is very fine with four layers of coils made by a chemical additive process, and the total thickness is only 0.7 mm.

6.7 Linear motors

This book has mainly been concerned with rotary motors, although ultrasonic-wave linear motors were mentioned in Chapter 1. We have said that few academics specialize in mass-produced conventional DC motors. In contrast, there are many scholars working on linear motors, even though there

Fig. 6.10 Three-phase linear stepping motor designed as DC servomotor. (Courtesy of Matsushita Electric Industrial Co., Ltd.).

is little demand for them. It seems that the linear motor brings to researchers a dream of future development.

Linear motors can be classified like rotary motors, a detailed classification of which is available in Appendix I. However, from a practical point of view, only limited types of linear motor are used.

6.7.1 Linear stepping motors

In the area of small precision motors, the linear stepping motor is a practical choice. Figure 6.10 is an example of one designed to be used as a servomotor. A position sensor is placed on the carriage and feeds back information to the electronic control stage, which decides which phases to excite. This is therefore a linear brushless DC motor, though it has the appearance of a stepping motor.

6.7.2 Voice-coil motors

Another important linear motor is the voice-coil motor shown in Fig. 6.11. Clearly, this may also be regarded as a moving-coil DC motor. The voice coil was originally utilized as part of a loudspeaker. The coil, to which the horn is coupled, vibrates at an audio frequency as a result of the interaction between a magnetic field and the current flowing in the coil.

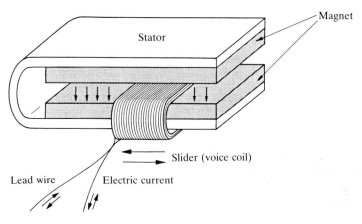

Fig. 6.11   Principle of voice-coil motor.

As is obvious, here a coil needs only to move left and right within a limited range in a magnetic field supplied by the permanent magnet. It is also obvious that, although this is a DC motor, brushes and a commutator are not needed. The voice-coil motor is used for the positioning actuator of a magnetic read/write head in a hard disk drive. As its speed can be higher than that of a stepping motor, it is suited to high-speed access demands in personal computers. Since a position sensor is needed with this motor, the cost is more than that of a stepping motor, which is normally operated in an open-loop control configuration requiring no position feedback.

6.7.3 Maglev bullet train

The most exotic linear motor is the driving force for the super-express bullet train which will run between Tokyo and Osaka or New York and Washington within one hour at a speed of 500 km h^{-1}. The motor will be synchronous with a strong magnetic field produced by superconducting coils built into the carriage as shown in Fig. 6.12. The armature windings are placed along the railway line. The trains will be levitated by the strong magnetic field to a height of 10 cm above the ground. Switching of thousands of amperes in the armature windings will be implemented by GTOs. This linear train system is called Maglev standing for 'magnetic levitation'. Maglev will consume huge amounts of electric power, but much less than aeroplanes. It is expected that, in the interests of saving energy, a Maglev bullet train network will stretch across North America in the future.

In the vicinity of the rails, there will be an extraordinarily large magnetic field that has never been created before. What effect will this have on living things? This is a new problem that will confront scientists.

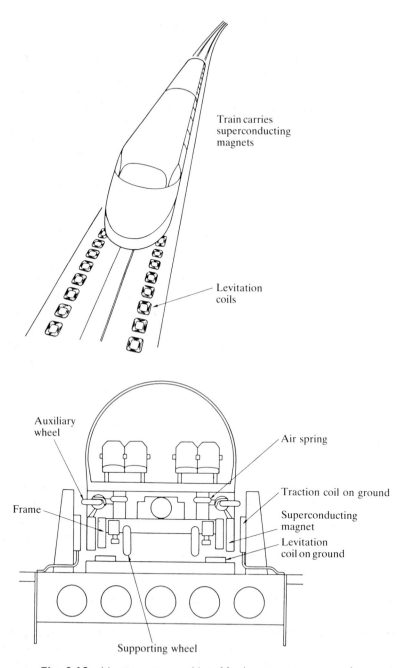

Train carries
superconducting
magnets

Levitation
coils

Auxiliary
wheel

Air spring

Frame

Traction coil on ground

Superconducting
magnet

Levitation
coil on ground

Supporting wheel

Fig. 6.12 Linear motor used in a Maglev super-express train

6.8 High-speed and direct-drive motors

Generally, electromagnetic motors are not suited to low-speed operation because of low efficiency. As discussed in Chapter 2, back emf plays an important role in pursuing economical electric-to-mechanical energy conversion. Since back emf is proportional to the rotor speed, the difference between the applied voltage and the back emf is large at low speeds and a large current flows. This means that much electric power is wasted for a low-speed low-output operation; most of the input power is dissipated in heat loss in the armature windings. Electric power is most efficiently converted to mechanical output at a speed close to no-load speed in conventional and brushless DC motors.

It is known that mechanical gears can reduce the speed and increase torque. A mechanism of belt and pulleys has similar characteristics. However, use of a gear or belt and pulleys degrades positioning accuracy and increases the size.

There are also several problems in high-speed designs. First, loss in the iron core due to high-frequency interaction between the magnetic field and the conductive material increases. Second, the bearing mechanism is a problem. The third problem is that a motor designed to be rated at a high speed is not suited to operate at a low speed because low-voltage high-current operation must be carried out. This means the loss in the control circuit also increases. Motors with permanent magnets are generally not suited to a wide range of operations.

There are big demands at the two extremes: high-speed applications and direct-drive low-speed applications. Typical requirements for the high-speed needs are: tooth grinder, centrifuges for uranium concentration, polygonal-mirror operation in a laser printer, and grinding machines. On the other hand, the requirement for direct-drive operation are: torso and arm operation of a robot.

For the direct-drive requirements, there are three choices: hybrid stepping-motor structure, variable-reluctance motor, and a brushless DC motor with high-energy magnets. Figure 6.13 shows a hybrid stepping-motor construction. Note that very fine teeth are cut in the cores. More studies are required to attain the exacting requirements of users.

6.9 The toroidal coil motor: an old motor in a new age

The DC motor invented by Italian physicist A. Pacinotti in 1860 and improved by Belgian physicist Z. T. Gramme had many commutator segments and a toroidal coil (see Fig. 6.14(a)). Though the design was found to be practical, it was not followed by any appreciable advancement. The main reason may be that only part of the coil was utilized for interaction with the magnetic field

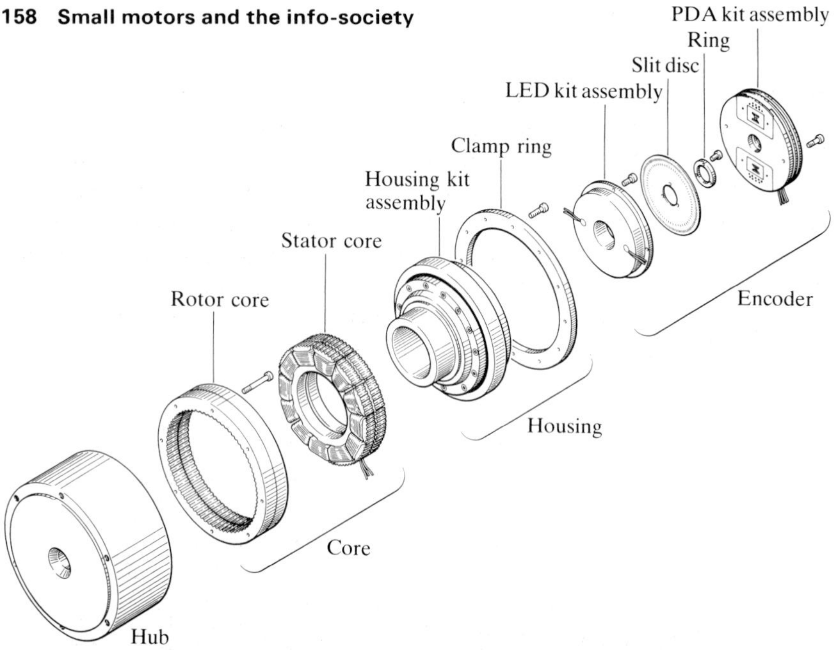

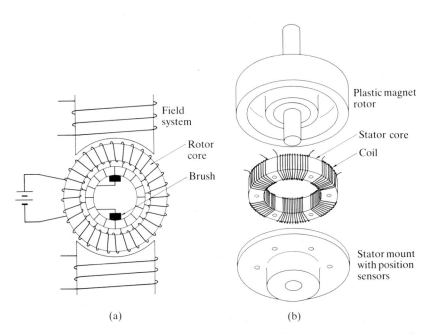

Fig. 6.13   A large hybrid stepping motor designed as a direct-drive servomotor (Courtesy Yokogawa Electric Corporation).

Fig. 6.14   Toroidal-coil motor; (a) principle of the DC commutator motor designed by Pacinotti and Gramme in the 1860s; (b) latest brushless version using a plastic magnet for the rotor.

and the armature or rotor construction was not always suitable for rotation. However, in modern times, this motor has been re-evaluated.

When this basic winding construction is utilized as a brushless DC motor, the armature is stationary and the magnet is the revolving component. The armature's three-phase winding consists of six coils in the motor as shown in Fig. 6.14(b). If we use a plastic magnet, shaped as shown in the figure which can cover three sides of the toroidal coils, the winding utility is better than that of normal motors. The design with single-layer coils is suited to high-speed operation (e.g. for a polygonal-mirror drive in a laser printer), because

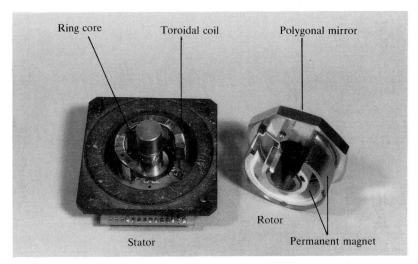

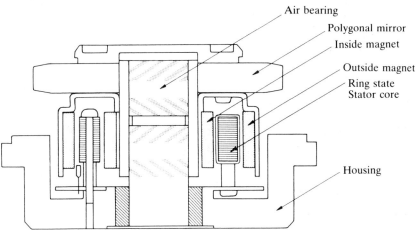

Fig. 6.15   Brushless DC motor employing a toroidal-coil armature and air bearings to turn a polygonal mirror at a speed of 400 rps (Courtesy Manufacturing Engineering Research Laboratory, Fuji Xerox Co., Ltd.).

the back emf generated from a small number of turns balances the low applied voltage at high speeds.

Though simple machine construction is advantageous, coiling takes much longer than with normal motors. If a chemical processing method could be used for this three-dimensional scheme, the toroidal brushless motor would be more widely accepted.

Look at the bearing construction in Fig. 6.15. Instead of ball bearings, an air bearing is employed in this motor. When the motor rotates, the shallow grooves on the shaft take in air to form a thin air film between the stator hub and the rotor shaft. As there is no friction in this bearing, the motor can turn at 400 or 500 rps without any problem. The biggest loss associated with this structure is the windage loss in the polygonal mirror driven by this motor.

6.10 A brushless DC motor without permanent magnets

The advantage of brushless motors in terms of reliability and absence of commutator and brush maintenance, which was stated in Chapter 3, applies not only to permanent-magnet motors but extends equally to a class of motors having no permanent magnets in the rotor. This sort of motor, known as the switched reluctance (or SR) motor, has only received serious consideration in the last 20 years, initially largely through the efforts of two British universities, and more recently on a commercial basis.

The principle of operation of the motor is probably the simplest of any, that of magnetic attraction: the basic function of a stepping motor without permanent magnets is dealt with in Section 2.2. The production of torque by magnetic attraction is apparent from Fig. 6.16 which shows a three-phase motor having four salient poles or teeth on its rotor and six poles on the stator with coils around each stator pole. A phase winding consists of the coils around the diameterically opposite stator teeth connected in series or parallel. As explained in Chapter 2 referring to Fig. 2.7 (p. 21), if each of these phases is energized in a counterclockwise sequence, A, B, C, the rotor will move clockwise, and vice versa.

Again look at Figs. 3.11 and 3.12 (pp. 60 and 61) to compare the similarities and differences of the rotors (3) and (7), both of which have salient poles. The former is the rotor of a reluctance motor, where the squirrel-cage conductors are needed to start and accelerate the rotor and the rotor's salient poles are essential for producing a torque at the synchronous speed. Note that the latter rotor cannot start normally on the 50/60 Hz mains supply. Brushless DC operation is a suitable means for providing this machine with starting and acceleration capabilities: this is the switched reluctance drive.

It should be noted that the direction of phase current is immaterial in the SR drive. This means that unidirectional current can be used, offering a very

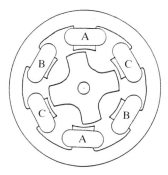

Fig. 6.16   A switched reluctance motor with six stator poles and four rotor poles, having three phase windings A, B, and C.

simple power circuit as compared with AC motors or permanent-magnet brushless motors, which require winding current alternating in direction together with complex PWM techniques. Moreover, because of the very high energy conversion capability of the arrangement when using a core of high magnet permeability, the torque that can be produced within a given volume substantially exceeds that produced in conventional machines where torque is created by magnetic action on the current or magnet in the rotor.

The switched reluctance motor, like the permanent-magnet brushless DC motor, relies totally on electronically switched supplies to its phase windings, timed to correspond with rotor position. It is clear that a pair of stator poles must be excited while a pair of rotor teeth is approaching. By synchronizing the SR motor's winding voltages to its rotor angular position, the production of torque and the efficiency are maximized.

Switched reluctance drives are competitive with other forms of variable-speed drives, both AC and DC over a wide range of power; they offer a higher efficiency than AC drives over a wide speed range as well as higher torque-to-volume ratio. At low powers they offer twice to four times the torque-to-volume output available from an induction motor.

The availability of a very wide speed range with the SR drive is due to the machine scheme having no permanent magnets. In a motor having a permanent magnet providing a constant magnetic flux, the back e.m.f. increases proportionately with speed. The speed is restricted to the level where the back e.m.f. balances the maximum supplied voltage. To be free from a permanent magnet implies a very high maximum speed.

Examples of three SR rotors, the two larger ones being for 8/6 pole, 4-phase motors, are shown in Fig. 6.17; the extreme simplicity and robustness is self-evident. Many other combinations of phases and poles are possible and can be selected as appropriate for specific applications. The present range of successful applications for SR drives extends from 10 W to in excess of 100 kW.

Fig. 6.17   Three typical SR motor rotors having ratings of 50 kW at 12.5 rps, 1 kW at 25 rps and 7 W at 167 rps.

and with rated speeds from several revolution per second (rps) to 50 rps with much higher speeds readily possible.

In fast response servo-type applications, the SR motor has the unique benefit of an exceptionally high torque-to-inertia ratio – by virtue of both a high ratio of torque-to-motor volume, and a naturally low inertia of rotor – together with a very wide torque bandwidth. As an example, an industrial SR motor rated 11 kW at 25 rps and not specifically designed or constructed for servo performance, takes only 100 ms for speed reversal from 30 rps to -30 rps, exhibiting acceleration of 5000 rad/s^2. Another example demonstrating the servo performance of a switched reluctance motor is its use by Hewlett Packard for one of their large A0-size plotters for very quick drawing.

Dr Rex Davis, Nottingham University, claims that these drives will be commonplace in domestic, office, and industrial applications before the beginning of the 21st century.

6.11 Frontiers in ultrasonic-wave motor technology

Toshiiku Sashida, the inventor of the ultrasonic-wave motor and a very humble engineer, commented to the present author:

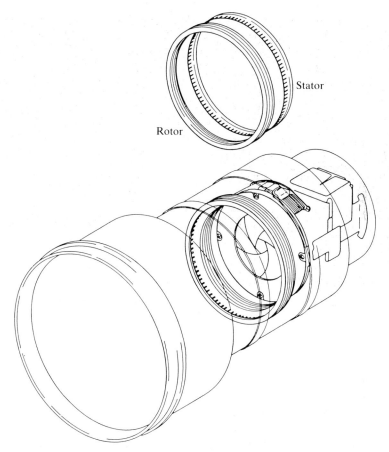

Stator

Rotor

Fig. 6.18   Camera with autofocusing mechanism using an ultrasonic-wave motor.

The ultrasonic motor cannot always compete with the conventional electro-magnetic motor in terms of overall performance and cost. However, the new motor has splendid characteristics: it is compact in size, generates a high torque at a low speed, and provides quiet operation. Insensitivity to magnetic field is the biggest advantage. Ultrasonic motors are suited to the application areas absolutely requiring these characteristics.

Autofocusing of a camera is one example. As seen in Fig. 6.18, the ring-shaped motor is compact and does not need a reduction-gear mechanism. The speed of the thin ring motor ensures that the focusing is exact even when several exposures are made in very rapid succession (e.g. 3 exposures/second).

(a)

(b)

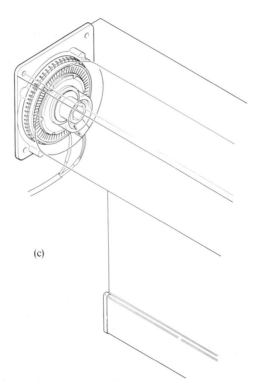
(c)

Fig. 6.19   (a) The new Tokyo Metropolitan office building; (b) one of the automatic roll-curtains; (c) the microprocessor-controlled ultrasonic motor is fitted to the top of the curtain.

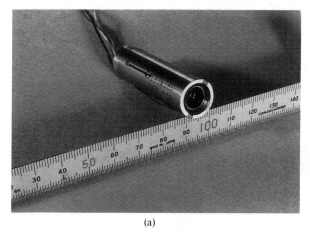

(a)

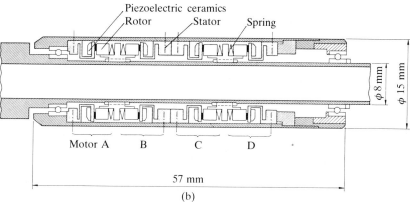

(b)

Fig. 6.20 Ultrasonic-wave motor designed for use in a laser-beam welding device in a nuclear power station: (a) housing and stator; (b) cross-sectional diagram.

More than a quarter of the 4000 windows in the new Tokyo Metropolitan office building use a roll-curtain driven by an ultrasonic motor as shown in Fig. 6.19, with the expected low-noise operation. Before installation a number of tests were done to ensure reliable operation of the microprocessor-controlled motors.

The ultrasonic motor is very suitable as the actuator in nuclear magnetic resonance medical equipment, because it is not influenced by strong magnetic fields. The new motor is also expected to operate without failure in the huge magnetic fields created by the superconducting magnet in Maglev trains.

One of the most suitable applications is as an actuator that works in a nuclear power station by remote control. Since the space allocated to the actuator is narrow, the actuator must be very compact. However, a high torque is required that cannot be attained by the conventional electrodynamic principle. The major reason for selecting an ultrasonic motor for this use is that not only does it exert a high torque even without a gearhead but it can also be constructed hollow so that other necessary things can be installed in the hollow space. Figure 6.20 shows a photograph and cutaway view of the latest motor design for use in a laser-beam welding device operated in a thin metal pipe. The hollow space is for glass fibre carrying a high-energy laser beam. The motor is used to operate the welding head to repair damage to the pipe. Mr Sashida is very busy designing these special-use motors with his engineers.

6.12   Analogy with biomechanisms

The movement of fish in an aquarium is always surprising. The fins comprise a large number of muscles and tiny bones, and each of them moves in a different way, producing a beautiful and effective overall movement. The control signals for those muscles are transmitted from the brain via a nerve network.

When considering the application of motors, the movement of living things is instructive. Motor control in the future will probably aim at the complicated but harmonious movement of every part of the machine like a living thing. To this aim, not only the study of biomechanisms but also the development of intelligent control techniques, including smartpower ICs, must be enhanced. The study of both actuators and control technology are necessary.

We present here a summary of an article[2] on the smallest motor on Earth:

> The mechanism which turns the flagellum of bacteria is interesting. The flagellum is like a sperm tail and functions as a propeller to swim in water. The flagellum's turning speed is about 100 rps, and seems to be able to reverse freely. The mechanism of the flagellum is believed to operate as follows (see Fig. 6.21). The hook is a free connector for changing the direction of the flagellum, the ring in the outer film is a frictionless bearing, and the one in the inner film seems to generate rotation. The rotary movement is generated only when hydrogen ions (H^+) go between rings S and M along the protein-forming linear rows. The protein row in ring S is shown by a white segment and that in ring M by the black segment. The flagellum's rotation occurs when the ions drift along the gradient of the ion concentration in the direction indicated by the arrowhead, connecting the protein stripes of both rings. However, this cannot explain how reverse rotation is achieved.

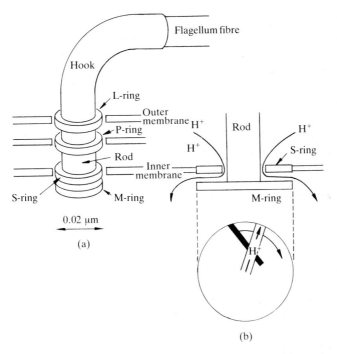

Fig. 6.21   The smallest motor on Earth: (a) flagellum root of bacteria; (b) flagellum motor.

   That this mechanism could be understood in the near future, and a new artificial motor working on a new principle could be created may not be a dream.

Conclusions

The value of the motor industry to our society was investigated from various angles, and frontiers in motor science and engineering were surveyed. Conventional AC induction motors and DC commutator motors will continue to be mass produced in the future. Some old types are seen to have been revived. Improvements will continue in the area of electronically controlled motors. New possibilities are being pursued in the area of ultrasonic-wave motors and actuators. Much is left in the study of biomechanisms for creating new micromotors and their controls .

References

1 Miller, T. (1989), *Brushless permanent-magnet and reluctance motor drives.* Oxford Science Publications.
2 Harayama, S. (1979), Revolving mechanism of bacteria's flagella [in Japanese]. *Shizen*, June issue, 26–31.

Appendix I
Classification of motors

There are several possible ways of classifying electric motors. Here we present a scheme suitable for small motors.

I. Classification by movement types
  [A] Rotary motors
  [B] Linear motors
There are more types of rotary motor than there are of linear motor.

II. Classification of rotary motors
[A] Motors having brushes and commutator
  ⟨A1⟩ Motors driven by DC current
    A1.1 Motors with permanent magnets
      A1.1.1 Slotted type, which uses iron core with slots
      A1.1.2 Slotless type
      A1.1.3 Moving-coil and coreless motors
        (1) Moving-coil motor, which places the field magnet outside the rotor
        (2) Coreless motor, which places the magnet in the hollow space of the rotor windings
      A1.1.4 Flat motor
        (1) Printed-circuit motor, whose windings are formed by stamping and welding
        (2) Pancake motor, which uses normal magnet wire
    A1.2 Motors using electromagnets
      A1.2.1 Series type
      A1.2.2 Shunt type
      A1.2.3 Separately excited type
  ⟨A2⟩ Motors driven by AC current
    A2.1 Universal motor (AC commutator motor, AC series motor)
    A2.2 Repulsion motor*
    A2.3 Schrage motor**

[B] Motors with AC motor construction no commutator
  ⟨B1⟩ Motors driven by AC current
    B1.1 Synchronous motors

B1.1.1 Hysteresis motor
B1.1.2 Reluctance motor
B1.1.3 Wound-field motor
B1.1.4 Permanent-magnet motor
B1.2 Asynchronous motors (induction motors)
B1.2.1 Squirrel-cage induction motor
B1.2.2 Eddy-current motor
(There are further three-phase motors, two-phase motors, and single-phase types of both synchronous and asynchronous types.)
⟨B2⟩ Motors to be driven by DC and an electronic drive circuit
B2.1 Brushless DC motor, which uses permanent magnets
B2.2 Switched reluctance motor, which uses no permanent magnets

[C] Motors with a stepping-motor construction
⟨C1⟩ Stepping motors
C1.1 Variable-reluctance type
C1.2 Permanent-magnet type
C1.3 Hybrid type
C1.4 Claw-tooth type
C1.5 Disc-magnet type
⟨C2⟩ Ultra-low speed synchronous motor, which has the hybrid stepping-motor structure but is driven on a single-phase source.

[D] Motors using piezoelectric ceramics
⟨D1⟩ Ceramic actuators
⟨D2⟩ Ultrasonic-wave motors

* The repulsion motor was not dealt with in the main text. An experiment can be done using rotor (8) of Fig. 3.11 inserted in stator B of Fig. 3.3. The brush assembly shown in 3.40 is used with the two brushes shorted with a wire. If a single-phase voltage is applied in the two-pole scheme of Fig. 3.8, the rotor will revolve. There is a suitable position for the brushes for each rotational direction.
** The Schrage motor was not explained, either. This is a unique three-phase motor, which can be operated over a wide speed-range around the synchronous speed.

Appendix II
History of motor science

(a) *Physicists and inventors who made great contributions to the development of electric motors*

Michael Faraday (1791–1867), English scientist. Born the son of a blacksmith, Faraday's schooling was very elementary. However, he made great contributions to the development of electrochemistry and electromagnetic science. His electrical experiments began to attract attention in 1821, when he demonstrated electromagnetic rotation. In this experiment, he showed that the flow of an electric current would cause a magnet to revolve around a wire carrying a current or cause a wire carrying a current to revolve around the magnet. The next ten years were spent trying to convert magnetic forces into electric forces, leading to the discovery of the principle of the transformer and generator.

Werner von Siemens (1816–1892) and Sir William Siemens (Karl Wilhelm Siemens) (1823–1883), German brothers, inventors, and industrialists. They made a lot of inventions. The self-excitation principle of the modern dynamo seems to have been first discovered by Werner, but it was William who published the results of their research in 1867. William became a British citizen in 1859, elected fellow of the Royal Society in 1862, and was knighted in 1883.

Nikola Tesla (1856–1943), American inventor. Born in Yugoslavia, and studied at the Polytechnic School in Graz, Austria. In 1884, he emigrated to the USA, and joined Thomas Edison's company. As he was interested in creating an AC motor, he severed his connection with Edison and established his own laboratory in New York City. In 1888, he invented an AC power transmission system and an induction motor supplied by the system. It is said that the 60 Hz transmission in North America originated with him. His name has been adopted as the SI unit of magnetic flux density.

Charles P. Steinmetz (1865–1923), American scientist and engineer. In 1889, to escape arrest as a political criminal in Germany, he fled to the USA and found employment as a draftsman. The electrical industry was then in its infancy, and Thomas A. Edison was committed to the development of a DC power system. Steinmetz became interested in the AC system. In 1893, he joined the newly organized General Electric Company and contributed to the development of the American electrical industry. The modern power system owes its existence to his work in opposition to Edison. His most important research was on the magnetic hysteresis phenomenon, and creation of AC motors with low loss associated with hysteresis.

Toshiiku Sashida (1939–), Japanese inventor. In 1982, he invented the ultrasonic-wave motor, utilizing the elliptical movement on the surface of an elastic substance vibrating at an ultrasonic frequency excited by piezoelectric ceramics.

(b) *Chronology*

1791   Michael Faraday was born in London.
1820   Dominique F. J. Arago (France) first experimented with magnetization from an electric current.
1820   Hans C. Oersted (Denmark) found that a force acts on a magnetic needle placed near a wire carrying a current. His name is adopted as the cgs unit of magnetic field intensity.
1820   André M. Ampère (France) established the law named after him to explain the discovery by Oersted. His name is adopted as the SI unit of electric current.
1821   Faraday demonstrated electromagnetic rotation.
1824   Dominique F. J. Arago experimented with a rotary magnetic field.
1831   Joseph Henry (USA) discovered the self-inductance phenomenon from his studies of electromagnets. He is thought of as the discoverer of the principle of modern motors.
1831   Faraday discovered electromagnetic induction.
1833   William Ritchie in London built a motor in which an electromagnet rotated around a vertical axis.
1834   Moritz Herman von Jacobi in Russia built a motor with a commutator.
1836   Thomas Davenport in the USA built a DC motor and drove a lathing machine.
1838   Robert Davidson in Britain built a DC motor and for a few years demonstrated its utility in applications in lathing machines, printing machines, locomotives, etc. At that time, steam engines began to be used in locomotives.
1860   Antonio Pacinotti (Italy) invented a ring-type armature with special features that permitted a device to be operated either as a motor or generator.
1866   Werner von Siemens and William Siemens invented the modern dynamo (generator).
1867   Zenobe T. Gramme (Belgium) made an AC generator, modifying the invention of Pacinotti. Faraday died.

1872   An AC generator was manufactured for the power source of the arc lamp. Thus, 40 years elapsed from Faraday's discovery to commercial utilization.

1873   James C. Maxwell (Britain) wrote *Treatise on electricity and magnetism*, from which stemmed many of the most significant discoveries of the twentieth century. His name has been adopted as the SI unit of magnetic flux.

1879   Maxwell died, and Albert Einstein was born in Germany.

1882   Nikola Tesla was inspired with the idea of two-phase AC motor, when he was in Budapest.

1885   Galileo Ferraris (Italy) discovered the principle of the rotary magnetic field, which made possible the development of polyphase motors.

1887   Tesla built a two-phase generator to drive his two-phase induction motor.

1888   Westinghouse bought Tesla's patents on the two-phase induction motor. Tesla built the first single-phase induction motor.

1889   Dolivo von Dobrovolski in Germany built the first three-phase squirrel-cage induction motor.

1893   Charles P. Steinmetz joined the General Electric Company.

1919   C. L. Walker (Britain) invented the stepping-motor structure to increase rotational resolution. However, it was 30 years after this that it was used on a commercial basis.

1937   B. R. Teare released the theory and his experiments on the hysteresis motor.

1948   The transistor was invented in the USA.

1952   K. M. Feiertag and J. T. Donahoo invented the hybrid stepping motor in the USA.

1957   VR stepping motors began to be used in numerically controlled machines in the USA.

1966   J. Henry-Baudot (France) invented the printed-circuit motor.

1971   The first microprocessor appeared in the USA.

1981   Claude Oudet (France) invented the disc-rotor stepping motor.

1982   Toshiiku Sashida (Japan) invented the ultrasonic-wave motor.

1986   The first product family of servomotors using neodmymium-iron-boron magnet appeared in the USA.

Epilogue

As a Sherlock Holmes fan, I felt I knew London even before my first visit in 1972. The occasion was the Conference on Electrical Variable-Speed Drives hosted by the Institution of Electrical Engineers. I stayed at a hotel in Charing Cross that I had read about in *The Hound of the Baskervilles*, and fortunately this place was not far from the conference and Baker Street.

The Institution building, standing by the River Thames, impressed me with the marble floor and pillars in its entrance hall. There, in the middle, was a statue of the great physicist Faraday, who, in 1831, discovered the physical law that has most dramatically improved our lives: electromagnetic induction. Beside it was a bust of Maxwell, the greatest theoretical physicist the nineteenth century produced.

Many of the motors invented, based upon the theories of these men, were displayed in the surrounding rooms when the Conference on Small Electrical Machines was held in 1976, and at that time many of the heated discussions in the plush auditorium were focused on the basic problems of stepping-motor design. Regrettably, I was not able to follow directly in English, but I was sufficiently inspired to study stepping motors and to write a book. After starting this work back in Tokyo, I became very interested in the motor's history and, with the help of Dr A. Hughes, University of Leeds, collected quite a bit of material on the subject. Eventually I discovered that two drastic improvements in stepping-motor design were made in Britain in 1919 and 1920. However, one of these was not used until a Japanese manufacturer began implementing it in designs for numerically controlled equipment some forty years later. It was in the 1950s in America that the other was combined with a permanent magnet to create the so-called hybrid stepping motor suited for precision positioning use. Nowadays most motors of this type are manufactured in Japan.

After I had completed the book in Japanese and it was published, stepping-motor design and production technologies began to progress at a spectacular pace. Thousands of copies were read in industrial circles. Brushless DC motors also started to be mass produced on a large scale for use in business machines, and it was during this period that I wrote the English version: my first volume was published by Oxford University Press.

So I was able to see what the initial stages of Japan's small motor industrialization were like and how we helped to create the age of copious office automation motors. Naturally there were many success stories, and more than a few failures as well, but one success stands out in my mind as a good lesson in the Japanese way of starting a company.

As stated in the Preface, I joined Teac in 1964, but the next year I transferred to the University of Industrial Technology, then called the Institute of Vocational Training, to restart my academic career as a young lecturer. My first lecture was in 1966 to ten students about to start their final-year projects. I talked mainly about AC synchronous motors because in those days that was the type predominantly used in information equipment. I did have one German-made brushless DC motor among the samples, however, and after the lecture when most of the students had already left, one came up to me and said that he would like to study much more about this new DC motor. As it turned out, this was to decide his future.

As soon as he agreed to do his final-year project under me, I gave him a German paper about the brushless motor. The next morning, much to my surprise, he was ready to discuss the paper: he had spent the entire night looking up every word in a dictionary without a single minute's sleep! He continued to study my theory of electromagnetism for motor design, and later succeeded me in my former position at Teac. In 1973, he rented a small cottage in Kyoto to start his own company with three men who had also studied at the same university. This young entrepreneur is Shigenobu Nagamori, President of Nidec Corporation, and various other American and European firms. With him I wrote *Permanent-magnet and brushless DC motors*, published by Oxford University Press in 1985. Now, ninety per cent of the precision brushless motors used in hard disk drives are supplied by his companies.

Although Britain, as well as America, produced people who did great fundamental work in science, it is regrettable to see how little manufacturing they have done in small motor technology with special reference to computer peripherals. In 1981 in Tokyo, I started *Motortech Japan* with the slogan 'Small motors—the driving force in the high-tech info-society!' and it soon became the world's largest show/conference in the field. Japan is now the largest manufacturer, but, apart from the ultrasonic motor, she has not contributed to the fundamental science at all. Is Japanese industry really in as prosperous a position as it seems to be? This is certainly a matter for worry. South-eastern countries around Japan are studying from her as she did from the West, and small motor design is a prospective and attractive target.

The brilliant discoveries in science and great inventions in technology achieved by the Anglo-Saxon have always been admired. However, it may have been an oversight for the British to have focused their abilities only on basic theory rather than making a continued effort in quality control and mass-production technology. But, likewise, it is probably Japan's error to improve strictly upon European inventions and production technologies, seeking only immediate commercial benefit.

As you have read in this book, we have improved and refined conventional electric motors, but we need new epoch-making progress in ideal motor design. In this aim, cooperation between European intelligence and Japanese expertise will be absolutely necessary. I am convinced that the British, their cultural successors, and the Japanese can share each other's skills and knowledge to create better motor technology for the future.

After Sherlock Holmes had perished at the Reichenbach Falls, Sir Arthur Conan Doyle agreed to bring him back to life in response to his readers' requests. Holmes was not killed in the fight with Professor Moriarty after all: his ju-jitsu skill saved him in the end. I truly hope Britain can also save its lagging industry in the end and resurface as a leader in electric motor science and technology by applying the Japanese philosophy in manufacturing technologies as well as training expert engineers and technicians.

In closing, I accept the advice of Professor T. Murase, a friend and fellow Sherlockian and cite a passage from *His last bow*:

> Good old Watson! You are the one fixed point in a changing age. There's an east wind coming all the same, such a wind as never blew on England yet. It will be cold and bitter, Watson, and a good many of us may wither before its blast. But it's God's own wind none the less, and a cleaner, better, stronger land will lie in the sunshine when the storm has cleared. Start her up, Watson, for it's time that we were on our way ...

Index

BRAIN FARTS

EWW EDITION!

BOOK ILLUSTRATIONS AND COVER DESIGN BY DAVOR RATKOVIC

In ancient Egypt when they would mummify someone, they would pull out the brain through the nose.

Why do you think the mummies look so tense? Because they're all wound up! And, probably because their brains were sucked out of their heads through their noses. Yikes!

So, what is mummification? It's a way of preserving a human or an animal by preventing its body from decomposing. And, this process was by no means easy or for the faint of heart. But the Egyptians sure did this job well. Their mummies are still intact, even after thousands of years!

How was it done? First, they would remove the internal organs, which was done through a cut in the stomach. The brain was trickier to handle and it required some skill. A special hook almost 7 inches long was inserted into the nose and into the brain. The hook was likely made from some part of a plant. Part of the brain would have been wound up around the stick and pulled out through the nose. The rest would have been liquified and drained out. They would leave the heart in the body because it was the center of a person's being.

Benjamin Franklin once wrote an entire essay about farts.

It seems that, apart from being a genius, Benjamin Franklin had quite a sense of humor. In 1781, Franklin suddenly decided that no other topic was more worthy of his interest than farting. He named it "To the Royal Academy of Farting" or shortly "Fart Proudly."

Now, why would he do a thing like that? Supposedly at the time, Franklin thought the Royal Academy of Brussels was too concerned with impractical matters and wrote this satirical essay as a response. He wrote to the Royal Academy, urging them to discover some drug (medicine, in this case), which could be added to people's food, so that the discharged wind would not be as offensive, and instead, smell like perfume.

He recommended scientific testing of farts and discussed how different foods affect the smell! He ends the essay with a pun, letting the reader know that he didn't take himself (or farting!) too seriously. The essay was never sent to the Royal Academy, but he did print a few copies to share with his friends.

In ancient Rome, people used to buy urine from public bathrooms.

Ancient Rome is well known for the extensive public bath and toilet system within the city. However, all of the urine from these toilets needed to be collected and disposed of. Urine was a valuable commodity as it was used for cleaning wool and tanning leather, It was also made into toothpaste and mouthwash! Let that sink in for a moment.

The emperor of Rome started charging a "Urine Tax" to the people buying the urine. Although the "Urine Tax" was not popular, the money that was earned helped to bring the Roman Empire back to its feet after a civil war.

A carnival park in California mistakenly bought a dead body and used it as a prop for a long time.

In 1976, a camera crew member was adjusting the mannequins in the Pike Amusement Zone for a TV show. When he moved the last one, he had a sight to behold. One of the mannequin's arms broke off. And, oh my... it wasn't a mannequin at all! It was a real human body!

Eventually, they discovered that the corpse belonged to a man named Elmer McCurdy, who was a better mannequin than he had ever been a man. He was an old outlaw and a train robber, who was killed in a shootout with police over 65 years earlier. After his death in 1911, his body had been mummified and used as a prop for traveling carnivals and sideshows for over 60 years. The mummified body was eventually sold as a prop to Pike Amusement Zone. After his identity was discovered, McCurdy was laid to rest in Guthrie, Oklahoma.

Australia went to war with a bunch of birds...and LOST.

Don't mess with an emu, kids. An emu is that big bird that doesn't fly, similar to an ostrich. Australians learned that lesson the hard way.

So, what happened? The emu birds were a protected species up until about 1922. That was when they started to eat all the crops, destroying everything in their path.Understandably, the farmers were upset. First, they tried shooting the emus, killing hundreds. Despite this, the emu population kept growing. So, the farmers realized they needed help.

This is where the Australian military comes in. Under the command of Major G.P.W. Meredith, the military headed in to gun down more emus. When they started to shoot, the emus scattered about, minimizing their chances of getting shot. Another unsuccessful attack took place. And finally, Asutralians accepted defeat.

The emus won and continued to be a problem for farmers for several years. The emus continued to be hunted occasionally and many farmers ended up using barrier fencing to try and protect their crops.

In the 18th century, it was not uncommon for a person to be buried alive.

During this time in history, the death rate was so high that many times doctors were too busy to come see if a person was actually dead. Caregivers of the presumed dead person would notice lack of breathing or a pulse as a sign that the person was deceased. However, sometimes they got it wrong and the person was buried alive!

In order to try and alleviate the problem, extra caution was taken. People would shout in the ears of the dead, stick needles under their toenails and even try whipping them to make sure they were indeed deceased. Special types of "safety" coffins were even designed so that a person could ring a bell in case they were accidentally buried alive!

One of the most famous philosophers in the ancient world used to run around naked.

Let's dive into the honest, naked truth about Diogenes. He was most known for giving up all his possessions and walking around naked, peeing and pooping wherever he wanted. He even had the nerve to tell Alexander the Great: "Get out of my sunlight." He criticized the teachings of both Plato and Socrates.

Basically, he expressed his anger and dissatisfaction with the current state of the world by disobeying the social norms any way he could. He did things which were shocking, even by today's standards. No written texts were left by him, but we still know much about him through oral tradition. He tended to prefer the company of wild dogs over people. He sounds like a pretty grumpy guy!

Soccer balls used to be inflated pig bladders.

Many, many years ago, humans had to get creative when it came to making toys and forms of entertainment. Pig bladders were used in sports, namely soccer and rugby. The bladder would be inflated and tied with a knot.

Unfortunately, they were too easy to rupture or break open, so someone thought of putting a leather cover over it. This made the balls rounder and it also made them last longer. There you have it: the first ball!

Boston once experienced a devastating flood, not from water but from molasses.

The Great Molasses Flood of 1919 happened as a result of a metal tank containing 2.3 million liters of molasses bursting open. A wave of molasses poured onto the streets traveling at around 35 miles per hour. Due to molasses being thicker than water it destroyed many houses, injured over 150 people and killed 21.

The police and firefighters arrived quickly and started helping the survivors. For the next few days, they rummaged through the mess and debris, finding trapped horses and ending their misery, and also recovering bodies. One body was stuck so hard that it was removed four months later. Cleanup took many weeks and utilized salt water for washing and sand for absorbing the sticky mess. The smell of molasses lingered in the air and the waters of Boston Harbor were browned all summer.

The incident resulted in the state's first class action lawsuit and is thought to be one of the first steps toward improving the safety of private companies.

Once, a Pope had a dead body dug up and put on trial.

This infamous trial is known as Cadaver Synod. It took place in the year 897, when Pope Stephen VI had the previous Pope's (Pope Formosus) dead body dug up out of the ground. He then commanded people to take it to the courthouse.

The reason? To find the old pope guilty of some past political scandals. The dead pope was provided with a deacon to defend him, but even an earthquake in the middle of the trial didn't stop the dead man from being found guilty. He was stripped of his fancy garments, had his three fingers used for blessing people removed and was then thrown in the Tiber River.

The trial caused disorder and chaos which eventually led to the imprisonment and death of Pope Stephen VI. Sometimes it's better to just have to let things go.

A mousetrap created in the mid-1800's caught a mouse in 2016.

Collin Pullinger created his Perpetual Mousetrap and claimed it would last a lifetime. It turns out he was right! The mouse trap has been on display in the English Rural Life Museum. One night in 2016, a mouse got into the display attempting to build a nest. The activity triggered the see-saw mechanism of the trap and sadly, the mouse didn't survive. But Pullinger's claim for his lifetime mousetrap sure holds up!

In the late Stone Age it was a common practice to cut a hole in your skull.

You'd never believe it, but there is actually such a thing as Stone Age brain surgeons who treated the warriors. The procedure, which was called trepanation, consisted of a hole being drilled in the skull of a person. The reason? To ward off evil spirits.

Scientists claim that the brain surgery was also done in an effort to cure super painful headaches or to help someone who recently had a head injury. They had no idea what epilepsy was, so they treated that with trepanation as well, probably assuming that they were warding off evil spirits.

Would you believe it, some people actually survived these operations!

People used to work collecting dog poop off the streets and sell it.

If I told you that the secret ingredient to leather back in the Victorian era of England was dog poop, you might say I was barking mad! But back then, dog poop was used to purify leather and make it more pliable. Leather was a very common good, being used for horse harnesses, saddles, shoes, bags and book bindings.

Dog poop was known as "pure" and the people who scooped it off the streets were known as "pure finders." These "pure finders" would frequent the areas that were known to have many stray dogs and scoop poop into a bucket to sell later to leather tanners. Some of the finders wore gloves, but others found a glove too cumbersome to scoop and so they just scooped bare-handed!

The Aztecs used pee to clean wounds.

If you are a warrior, you know a thing or two about wounds, because odds are you have had many. After some trial and error, this is the treatment the Aztecs came up with: first, you pee on your wound (it has to be fresh and warm, no cheating!).

Urine was thought to be sterile (it's not), so they used this as a way to disinfect the wound. Second, you use an herb to treat the wound so the bleeding will stop, and finally, wrap the wound with hot sap from Agave leaves.

The Aztecs knew how to cure or fix most any illness or ailment, with one of their 3,000 medicinal herbs, some of which even served as anesthetics or were used in dental work. I'm really happy that doctors don't pee on their patients today.

Fake teeth used to be made with teeth pulled from the mouths of dead soldiers.

The beginning of the 19th century was still too early for proper dentistry. And, the rich, whose teeth were particularly rotten (because of all that sugar and attempts at teeth whitening with all kinds of acidic solutions), needed a new set.

So, where would they get it? Why, from the mouths of dead soldiers of course! Looters, scavengers and sometimes, even the surviving soldiers would pull teeth from the dead soldier to make a profit.

The base plate of the dentures was ivory, then the real teeth would be attached to it. That would cost around

$120 or even more, which was a small fortune back in those days. A cheaper option was to have an ivory base with ivory teeth as well.

It seemed these new teeth didn't last long either. Putting older teeth into an already unhealthy mouth was not a recipe for success.

Under the streets of Paris is a tunnel system decorated with the remains of over 6 million dead people.

During the 1700's Paris was so heavily populated, that the city ran out of places to bury their dead. When a cemetery wall collapsed, spilling corpses into the streets, something had to be done!

In order to free up some space for future burials, the remains of over 6 million people were excavated and brought to an underground tunnel system that ran below the city. The tunnels that were once limestone quarries, now are lined with intricate designs and symbols, all created out of bones. There are literally miles of walls lined with bones!

Today, about 1 mile of the Paris Catacombs are open to the public for touring.

One day in the middle ages a group of nuns in a French convent started meowing like a bunch of cats.

It all started innocently enough. One nun was obviously overwhelmed by hysteria or something of the sort, and she started meowing. After about a week, apparently all the nuns had taken up this strange hobby, and it is said that they would meow in each other's faces for hours on end. Eventually this went on for so long and became so annoying that guards had to threaten to beat up the nuns until they finally stopped.

In medical terms, this is called mass hysteria or collective delusion. And, the example of the meowing nuns is a perfect one. It is described as a particular behavior which is taken up by a lot of people. It usually happens in a close community and the occurrence spreads quickly. Luckily, it usually doesn't last long.

There was once a popular diet that had people swallowing tapeworms and other parasites.

Sometimes, reality is really stranger than fiction. And, the people of the Victorian era in England are a perfect example of that. This was the time period during the late 1800s. Science wasn't quite as advanced back then as it is today.

We all know someone who's at some point been on a diet. If done properly, it can actually benefit our mental and physical health. However, if one gets a little carried away, they might end up with tapeworms in their belly on purpose!

Definitely not for the squeamish, the tapeworm diet promised you could eat as much as you'd like and still keep losing weight. Sounds simple, right? But, also disgusting, because the tape worm could grow as long as 30 feet, and start wreaking havoc inside by causing headaches and eye issues, meningitis, etc. But, they believed beauty is pain, and they were willing to sacrifice their health for it. Crazy, right?

The Romans used human pee as mouthwash.

The ancient Romans were pretty clean people, which draws some concern for this fact.

According to research, the ancient Romans turned to urine when they needed mouthwash, toothpaste, or laundry detergent.

There is something called ammonia that is found in urine. Ammonia is a compound of nitrogen and hydrogen and is an odorless gas. Even today, many commercial cleaning products contain ammonia, so it turns out that the Roman's thinking wasn't too far out there. Ammonia certainly has disinfecting qualities and can make your teeth whiter. Plus, it prevents cavities. So despite the yuck factor, it seems urine would be exactly what you needed back in the day.

I think I'll just stick with mouthwash, thank you.

Without flies there wouldn't be any chocolate.

Ok, it's a bit more complicated than that, but here's the deal. There is a teeny tiny fly, or a chocolate midge, that is responsible for pollinating the cacao tree. The tiny bugs are the only ones that can work their way into the intricate flowers that grow on the tree's trunk. These midges are only about one to three millimeters long, and their tiny size means they are pretty much the only bug that can maneuver itself into the flower.

The midges have a big job, too. It is estimated that only 1 in 400-500 flowers will produce fruit and only 10-30% of those fruit "pods" will reach maturity and be able to be used in making chocolate!

In Japan, people eat a poisonous puffer fish...which can kill you.

100 people die every year from eating this poisonous fish. Yet people still eat it! Japanese chefs must train for many years to learn the technique to prepare a puffer fish for human consumption so they don't KILL their customers.

That's because the pufferfish contains tetrodotoxin or TTX, a chemical that is 1200 times more potent than cyanide. This chemical can cause muscle fatigue, paralysis and even death. The highly poisonous liver is considered a delicacy in Japan, but other parts are actually illegal to prepare and eat.

Chefs need to study and train for three or more years and take several exams to prove that they are capable of correctly preparing this dangerous dinner!

One rodent hair is allowed to be in every 100 grams of peanut butter.

Yum! The FDA is the USA's Food and Drug Administration. And they say that it is simply impossible for food not to contain certain bits of animals, insects, poop, all kinds of stuff! And there's no doubt that there is a good bit of truth in that.

So if you were to make a peanut butter and jelly sandwich (one of my personal favorites), from the peanut butter alone, you can expect as much as 8 tiny insect fragments, along with a tiny bit of rodent filth like a piece of hair maybe.

It gets worse with the jelly...it's totally fine for there to be an average of 4 rodent hairs with 5 entire insects (not just parts) in every 100 grams of a product like apple butter. It's also fine for as much as 12% of it to be moldy. Other jellies can be 30% or as much as 75% moldy!

Are jelly beans and candy corn made from bug poop?

Shellac, which is used to give candy such as jelly beans and candy corn their shine, is made using excrement secretions from the female Kerria lacca insect. In other words, that candy coating really is made from bug poop!

Whoah! Yes, this insect is a beetle that lives in Thailand and India. In India alone, 20,000 tons of this bug's poop is collected each year. How would you like to work in a bug poop factory?

So if you want to tell your friends that candy is actually made from bug poop, you won't be entirely wrong. The bug's poop is scraped off of tree branches in the forest and used to coat your tasty candy. This is just another good reason to brush your teeth!

You really can die from eating too many apple seeds.

We all know that an apple a day keeps the doctor away. But, what if I told you that there's something in that same apple that might kill you?

Your average apple has between 5 to 8 seeds. Each of those seeds is made of about 1 to 4 milligrams of amygdalin, a substance that when chewed and digested can release cyanide into the bloodstream. So, what is cyanide? It is a fast acting, potentially lethal chemical that prevents the body from using oxygen in the cells. Cyanide is actually one of the deadliest poisons!

But have no fear and continue to eat those apples, because it would actually take you carefully grinding at least 200 seeds (30- 40 apples) to cause any real damage. The pits or seeds from peaches, cherries and apricots also contain cyanide, so best to avoid eating those seeds, too.

Farm-raised salmon is dyed pink to look like ocean caught salmon.

Salmon is a tasty fish known for being rich in Omega-3 Fatty Acids and having a bright orangey-pink color. It has become so popular that it is being produced on salmon farms to keep up with demand.

But the fish raised on farms is actually grey and not naturally pink like wild caught salmon. This is because wild caught salmon eat shrimp and krill, and these shellfish contain the compound astaxanthin, which is responsible for the orange-pink color.

In order to "look" more appealing, salmon farmers often feed the fish a pelleted feed that has a dye in it. This way the farm raised fish will still look like its wild caught relatives and be more appealing in the grocery stores.

Wombats poop is cube shaped.

Everyone knows that wombats are super cute and pudgy. But, not many people know about the advanced wombat technology of square poop. Yes, you heard that right. Their poop is cube shaped, and they stack it up to mark their territory and communicate with other wombats around. Research on this rare occurrence in nature is still ongoing, but there is an explanation.

The shape of poop is regulated by the pressure inside the intestines, and namely, the shape of the intestines themselves. For example, a pig has uniform elasticity of its intestines, so its poop is more round in shape. Wombat intestines are more irregular, and there are two parts where it's a bit stretchier, and this is where the cubing takes place. I wonder if that would go down the toilet, though...

Kangaroos constantly lick their arms.

Sounds easier to just take a bath, right? But, kangaroos will only take a bath if it's utterly necessary, like if they've just run away from a predator. The rest of the time, they just stick to licking their arms to keep cool. A spit bath, if you will.

If you think about it, a kangaroo's thermoregulatory activities are pretty awesome: panting, sweating and licking.

So, basically, they're like dogs. Only, a kangaroo has a special network of superficial vessels on its forearms, which means they're very close to the skin. They leave a layer of saliva, and as it evaporates, a kangaroo's temperature cools down.

Next time you feel too hot, give it a try. But maybe not if there's a bunch of people around.

Sloths only poop once a week.

Sloths are totally chill and lazy. They are never in a rush to do anything. Sloths take their time. So it's no wonder that's the case with pooping, as well.

They poop rarely, only once a week, and always in the same spot. How polite of them, right? It's probably not to step in their own doody. This way, at least they know where the toilet is.

But, it's important to mention why this is so. A sloth's metabolism, like himself, is slow. Like, super slow-motion slow. A sloth has to climb down from the tree, where it is safe, and poop on the ground, where it's anything but safe.

Their weak hind-legs make it more difficult to move on the ground, so pooping makes it extremely vulnerable to predators. They could easily get eaten while pooping.

Just imagine pooping being one of the most dangerous things you could do.This is because, like everything else, it takes the sloth a while to poop. Not only because sloths are slow, but they can poop up to ⅓ of their body weight.

When you only go once a week, it all stores up!

Hagfish eat dead fish by eating them from the inside out.

If you've been wondering what a hagfish is, then you're not the only one. Basically, hagfish are these eel- shaped, tube-like fish who are totally blind, but they have these whiskers which they use to explore the world around them. So, what - you'd ask. Just some freaky fish, right?

Wrong. These are more than just freaky fish. These are super old, about 360,000,000 years old to be exact. These fish feast off carcasses on the seafloor by digging into them and then start eating not only with their mouth, but with their skin as well. They soak up the much needed nutrients and if they are ever attacked, they can shoot up to 17 pints of mucus at their attacker. That's a pretty crazy defense mechanism!

This might be THE grossest animal on the planet.

The blue whale's fart bubble is big enough to fit around a horse.

Just because some mammals live underwater, it doesn't mean that they don't pass gas. They do. It just doesn't smell as bad. It also doesn't blast off as bad, due to the water. But it still smells, according to scientists. Still, what's most interesting about underwater farts is the size of the fart bubble. In the case of our good friend the blue whale, it can be big enough to fit around an entire horse. This somehow doesn't come as too big a surprise, seeing that the blue whale is the largest creature on Earth.

Farting is a common response to the pent up air in our bodies. It is released in one of the two ways: through the mouth or through the booty. Due to their size, blue whales have massive digestive systems and they eat tons of food. Hence, massive, smelly, horse-sized farts. There. Now you know.

Ferrets are the easiest animal to clog due to their intestines being the size of shoelaces.

Ferrets are fast animals with equally fast metabolisms. Also, they are very curious about their surroundings and have a tendency to chew on just about anything.

Being that they are carnivores, they need a diet which is high in protein and low in carbs and sugars. Their stomach acids dissolve food quickly, in about three hours or so.

But because ferrets have long, narrow intestines, they can easily get blocked by some foreign object the ferret isn't supposed to be eating, such as rubber or plastic objects they find lying around or even by its own hairball!

The pacific barreleye fish can rotate its eyes upward and see out of the top of its head.

Looking more like an alien than an actual fish with that submarine head, the barreleye fish is the thing your nightmares are made of. Also known as the spookfish (gee, I wonder why?), it looks like one of Mother Nature's jokes, but in fact, there is a very good reason for that see-through head.

It lives pretty deep in the ocean, around 2,500 feet deep to be exact, and this weirdly shaped head is an ingenious hunting tool which helps the barreleye fish illuminate the darkness it lives in. It feeds on jellyfish, crustaceans and other small animals which can be found around it.

And when it wants to have some fun, it spooks them by rotating its eyes upward. Boo!

Most penguins throw up in their babies' mouths to feed them.

If someone asked you to think of the most unusual bird in the world, which one would it be? A penguin, of course. When they are born, they find themselves in colonies, which are big groups of penguins, who are very social and tend to stick together.

If a chick hatches while its mother is away (the father sits on the egg to keep it warm, while the mother gathers food), the father will regurgitate food into the chick's mouth. This is actually considered a special kind of milk, and not all penguins are capable of doing it.

Once mommy penguin comes back, she takes over, and dad can go look for food. She will also feed it regurgitated food she had collected during her time away.

Aren't you glad your parents don't feed you like that?

Lobsters pee out of their FACE.

A face is for many things, peeing not being one of them. Lobsters just didn't seem to get the memo. Let's see how this crazy thing works.

Apparently, lobsters have two urinary bladders, on both sides of the head. When they communicate, they use smells and the scents that are in the urine. So, when necessary (when they want to attract a mate or warn a rival to keep clear), they will pee, that is, project their urine about 3 to 6 ft in front of them.

I myself would be totally warned off if my rival peed out of his face, and I'd just step down. Getting peed on sounds pretty terrible to me, whether it's from a lobster face or not.

Sharks have to always move forward or they will get water in their gills and suffocate.

We all know sharks to be the kings of the water. It is their playground, their kitchen where they feast on whatever they want. But, can you imagine a shark ever swimming for its very life?

It's hard to believe, but it's true. A shark will drown if it stops moving, so it has to keep swimming at all times, and it has everything to do with the way it breathes. The water comes in the shark's mouth, and then flows over the gills.

The oxygen from the water is absorbed, and the rest just flows out of the gills. Now, because there is so little oxygen in the water itself, the shark constantly needs to take in water. Just imagine constantly needing to move in order to stay alive!

The only three birds that can produce milk are Pigeons, Flamingos, and Male Emperor Penguins.

Birds can't produce milk. They're not mammals. Right? Well, there is an exception to every rule, and in this case, those exceptions are pigeons, flamingos and male emperor penguins.

Pigeon milk isn't like cow's milk. It still contains all the immune-enhancing and antioxidant benefits, but it's not made the same way. Mammals produce milk through mammary glands and teats, but birds (both males and females) regurgitate their milk, and it's called crop milk.

Regurgitated milk. Yum, right?

However we feel about it, the baby birds love its cottage cheese consistency and it helps them grow up into big, happy and healthy birds.

Babirusa pigs have a set of teeth that grow up all the way through their snout.

The babirusa is often referred to as a wild pig with a dental problem. Their tusks or canine teeth grow backward, piercing right through the skin in the snout, finally curving towards their forehead. If you are having a hard time picturing that, imagine if instead of your toenails growing outwards they grew inwards and came out the back of your feet. Crazy, right?

Their name babirusa actually means pig deer in the Malay language, because they reminded people of deer antlers. However, the true purpose of these weird looking tusks is still a mystery. They are pretty fragile, so it's generally believed that they are just for show, signaling to the females that they are good potential mates.

Many frogs eat their skin after shedding it.

Have you heard about this? It's ribbiting, isn't it? Eating your own skin.

Why would a frog shed its skin in the first place? Simple: to keep it from hardening. Because frogs intake oxygen in the water through their skin, it slowly hardens. Once it becomes useless, the frog sheds it.

This happens very often, and it's painless. The frog just scrunches up, crouches, then rips the old skin and starts stretching out of it. Shortly after, the process is finished. A piece of cake.

However, it also causes the frog to lose a lot of nutrients. So, the frog just eats the skin (that is full of the missing nutrients) and voila. Problem solved.

Earthworms have five different hearts.

If you think of a heart in terms of the human heart, then the earthworm's got none. But, they do have something else, something equally important.

They have 5 pairs of aortic arches, which have more or less the same role as the human heart. They are spread along the central body cavity and make circulation of blood possible. Imagine the body of the earthworm as a rubber hose, and the aortic arches are the person holding the hose.

Every once in a while, that person squeezes the hose, and it is this contraction that sends the blood where it needs to go. So the next time you are feeling lonely, find an earthworm. They have lots of love to give.

Dogs lift their legs to pee so other dogs will think they're really big.

When a person pees, it's just for relief. But, when a dog pees, the whole thing is much more cryptic and complicated and wrapped in a veil of mystery. OK, maybe not a mystery, but have you ever wondered why a dog raises its leg high when peeing? I have. And, I've got the answer.

For a dog, peeing is more than just relieving himself. It's a message that says "I was here and you better recognize it." By marking the territory this way, a dog tells all other dogs about himself. It's like a personal letter written all in pee, stating what breed this dog is, whether it's stressed or not, overall health and size - you get the idea. And, whoever is peeing wants to do this right.

Also, the dog might even want to lie a little bit, just a little though, by raising his leg higher, to appear bigger and intimidate potential intruders.

Termites contribute to global warming because they fart so much.

If someone told you that the number one farters in the world were termites, you'd be like, no way! But, actually, it's true! Humans, cows, and elephants fart, and they fart a lot. But, it's nowhere near termite farts.

You all know that one of the main causes of global warming is methane. Methane can be found in farts. And termite farts seem to be the deadliest for our ecosystem. Termites are considered the second largest natural source of methane emissions. They must fart a lot, huh? That they do. Methane is produced in their digestive system, and these emissions depend on several factors, such as population of insects, their species, and so on.

Bears can eat 40,000 moths a day.

If you go out in the woods of Yellowstone National Park in Wyoming, you might be in for a big surprise if you stumble upon bears having a picnic. What do you think they'll be eating? A delicious moth sandwich!

No, seriously. Each summer, moths burrow into dark crevices, hiding away from the sunlight. Then, grizzly bears and black bears come to feast. Imagine this: just one tiny little moth has enough fat to count for half a calorie. The bears can devour as many as 40,000 moths or 20,000 calories in a single day.

They all stand in line, at a safe distance from one another while feeding. Otherwise, they'd fight each other and there would be no delicious moth bites for any bear.

Some perfumes have samples of whale vomit in them.

When I'm looking for a good perfume, I like it to have citrus scents. Rare wood. Maybe some exotic flowers. But, I don't expect it to contain whale vomit. Well, it seems I'm asking too much.

If you take a look at any perfume label, you'll see there's something in it called ambergris. Sounds fancy, doesn't it? It's not! It's whale vomit.

If it's any consolation, only 1% of sperm whales can make viable ambergris. They expel it, and it's waxy at first. It hardens over time and eventually rises to the surface, ending up on the shore. Oh, and it smells like marine fecal matter. If you're looking for it on the beach, you have to really know what you're doing, because they resemble any old rock.

So, next time you spritz on some perfume, remember where it came from.

Cat tongues are made up of the same stuff your fingernails are.

Why do cat tongues feel like sandpaper? Simple. Because they're covered with tiny little claws.

What? I hear you ask. Tongues made of claws? Ridiculous. Well, not quite. Cat tongues feel so rough because they're covered in tiny hooks which are made of keratin, the same stuff that your fingernails and toenails are made of.

Now, why on earth would a cat need nails on its tongue? Again, simple. Cats can detangle any knot in their fur and remove any parasites lurking in there. Also, this way, they can redistribute oils all over their fur, which makes for proper waterproofing. This is why your vet says never to wash your cat. It does it all on its own with that little sandpaper tongue.

Platypus mothers sweat milk to feed their babies.

If I told you that a duck/beaver creature with the legs of an otter sweats milk, you'd probably think I've gone crazy. But actually, everything stated in the previous sentence is true. Honest!

Let me explain. Mammals, including humans, secrete milk. It comes from mammary glands and exits the female body via teats. Now, if you've ever seen a platypus, you might have noticed that it has no teats. So, the nipple in mouth method doesn't work here.

What platypus mothers do instead is they sweat out their milk, and the babies just lick it off their skin. It seems like sweat, but platypuses don't sweat, being aquatic mammals and all.

Also, seeing that this method is rather unsanitary, platypus milk is enriched with additional antibacterial proteins, so that the babies won't get sick.

Turtles have the ability to breathe through their butts.

And no, I don't mean fart. I mean cloacal respiration. But, let's take it from the top and explain this butt breathing properly.

Turtles in general can stay underwater for longer periods, even months at a time, especially during winter when the surface of the pond or lake is iced over. So, when they can't rise to the surface to ingest oxygen, what do they do? They breathe through their butts.

Well, technically, they respire. They don't breathe. And, this can only be done when they're hibernating, because their heart rate is slow, they don't move and their metabolism is super low, so they don't use up much of their stored oxygen.

Basically, the difference between breathing and butt breathing is that butt breathing can't provide oxygen for an active turtle, only a hibernating one.

Childbirth is painful for humans, other mammals...not so much.

Humans are completely different from other mammals. We walk on two legs, and our brains are much bigger. And, this is so right at the onset. Our babies are born with big, active brains, but with bodies that need a whole year to start moving properly and with focus.

Childbirth is no different. For human mothers, it lasts an average of nine hours. Apes, for example, do it in two hours. That is because an animal mother has wide hips and the head of her baby is much smaller than the head of a human baby. So, she gives birth easily and with less pain.

Turkey vultures defend themselves by puking into your face.

While they do resemble turkeys with their red heads, white bill, and dark brown feathery body, there is nothing even remotely cute about turkey vultures. They're scavengers, who don't rely on killing smaller animals for food, but rather feasting off of an already dead animal. This saves them the hassle of actually killing them. In a way, they're doing us all a favor, cleaning up the land of rotting carcasses.

If they get disturbed during their feeding, they have a very effective way of keeping other animals at bay. They regurgitate their food, and send it right into the face of their attacker. What's even more amazing is that they can project it 10 ft away, so whoever has got their eye on a turkey vulture's dinner, they better think twice, unless they want to end up with a face full of rotting, regurgitated flesh.

The viceroy caterpillar looks just like bird poop.

While a viceroy butterfly is a beautiful insect whose wings are the loveliest shade of red and orange, the viceroy caterpillar is a whole different story. Brownish green, with streaks of white, the viceroy caterpillar boasts a most ingenious method of camouflage. What is it? Drumroll...it looks like bird poop. Clever, right?

This way, looking like some dried up turd everyone would just frown upon, it makes sure that it is left alone and not eaten by predators.

However, if you take a closer look you'll see it's got some more oddities, like humps and bumps, and two black antennae. As if looking like poop isn't bad enough, they have to look like dried alien poop, too.

The most poisonous spider in the world can bite through your shoe.

You've all heard those horror stories about a friend of a friend who had a scary encounter with a spider, such as being bitten then falling into a coma, even dying, or spiders laying eggs inside their ears, and so on. Frightening stuff!

But what's even scarier is the fact that some of it might actually be true, and not just the stuff of urban legends. Enter Australia's funnel-web spider, the most venomous spider species of all. I'm not even joking. If someone gets bitten by this spider, whose teeth can actually bite through shoes and fingernails, that is considered a life-threatening emergency and is treated as such. Unfortunately for Australians, they are mostly found in Sydney, Queensland, South Australia and Tasmania.

Still, only about 14 deaths have ever been recorded. So, we can rest easy... until we see a spider and slowly start backing up. Some of us might even screech. Just a little.

A camel's pee comes out thicker than syrup.

Camels can go for days without water. Everyone knows that. Imagine going a week without a single drop of water, while walking the sands of the Sahara desert! You'd think this super awesome skill is because of their humps, right?

Wrong. While any other animal would die of dehydration, a camel keeps on going not because of its humps, which are just big mounds of fat, but actually because of their oval-shaped red blood cells (blood cells in other animals are circular) and their kidneys.

Now, you might go: big whoop, everyone's got kidneys. But not camel kidneys. Those babies are so efficient that they filter out most of the water to help keep the camel hydrated. Therefore, a camel's urine can be as thick as syrup!

There are 200 million insects on earth for every human.

You don't need a book to tell you that insects are the most diverse creatures on the face of the Earth. Their number also exceeds the number of any other species. Just look around you.

They're hiding below the ground, on the ground, inside every nook and cranny. They're basically everywhere. It is estimated that there are around 900,000 different kinds of insects in the world. That's a whopper of a number.

How come you might ask? Well, they've got incredible reproductive capabilities. Just take the east African termite queen. It lays an egg every two seconds. That means she has 43,000 eggs. Every. Single. Day.

No wonder then that there are 200 million insects for every human on this planet!

The average garden snail has over 14,000 teeth on its tongue.

Let this terrifying fact sink in: snails have thousands of teeth. On their tongue.

This tongue is pretty much like our human tongue, only it's got these little teeth. Their tongue is called the radula, and it is used to scrape off parts of their food as they're eating. And, because these teeth are not very strong, they get worn out very quickly and easily, so they are replaced pretty often by new ones.

Now, I know what you're thinking. Do snails bite? Yes. Yes, they do. But, don't be scared. A snail bite can't hurt you at all. Actually, it might even tickle a bit. The only thing all those teeth are good for, is nibbling on leaves and vegetables. Not human skin.

Prairie dogs touch front teeth to greet and recognize one another.

Despite what their name tells us, prairie dogs aren't dogs at all. They're actually a kind of a squirrel. As if that wasn't obvious just by looking at their cute little mugs, right?

Prairie dogs are very territorial, but that doesn't mean they aren't social. They're actually very social, living together in neighborhoods and communes, which are made of family groups. This is why it's so important to recognize one another as belonging to the group or not.

How do they do this? By kissing.

They lock their teeth together to see if the other one belongs to the same family group. If he does, then some playing and grooming will follow. If not, then they could become rather aggressive in an effort to defend their territory.

Ghost crabs have three teeth in their stomach that help grind food and make a growling noise.

Crabs don't like to share food. Why? Because they're shellfish! Haha!

All jokes aside, the ghost crab has three teeth in their stomach which means they can eat whatever they'd like. The crab has no teeth in their mouths, so the teeth in their stomachs are responsible for breaking down the food. This is called a gastric mill.

Now, because these three teeth actually grid against one another while working, it's no wonder then that they make a certain growling noise. This noise can also be heard when the crabs are aggressive, that is when they're trying to attack another creature with their claws. It also helps to ward off predators. And we thought our stomachs growled!

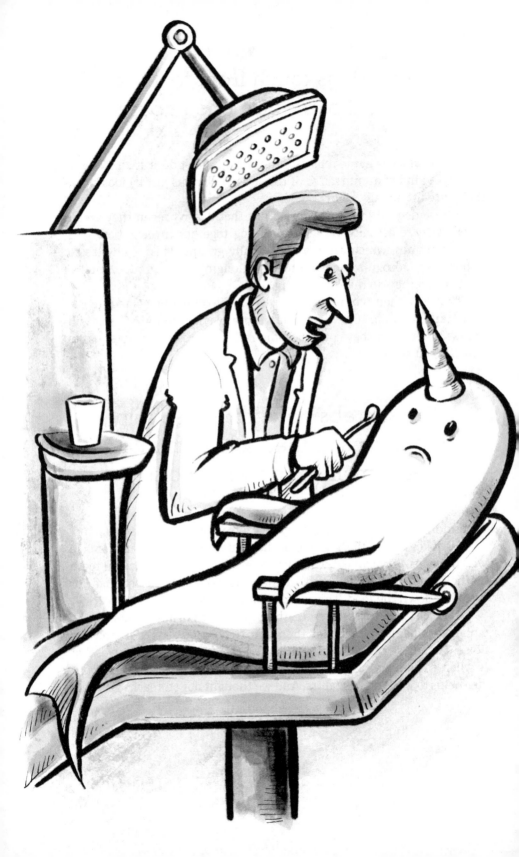

The narwhal's large horn is actually a giant tooth.

Unicorns aren't real. Everyone knows that sad truth. But, there's a cetacean that lives under water, which does have a long horn in the middle of its forehead. Only, that horn is really a tooth.

Narwhals have mystified explorers and researchers for centuries, and they still do. They have a long tooth, which on male narwhals can reach the length of approximately 9 feet. And, it's the only tooth they have, even though they belong to a group called toothed whales.

The purpose of this single tooth is still a mystery. Some say it's for defense or attack, others say that it's used for breaking ice, or even as a courting advantage. Whatever the real reason, narwhals remain shy and hidden from the human eye, only strengthening the mystery around their existence.

Manatees control their floating...by farting.

When Christopher Columbus said that he and the rest of the sailors saw mermaids during his trips, he was actually talking about manatees. Probably a trick of the sun, or maybe eating some bad food. But I suppose if he heard them fart, he'd know they definitely weren't mermaids.

Perhaps you've seen how gracefully manatees swim and float in the water. But have you seen them fart? You must have because they depend on flatulence to do all that. Releasing gas allows them to sink, while storing it makes them float. Taking into account that a grown up manatee eats about 100 to 150 pounds of vegetation on a daily basis, it's no wonder that they've got a methane bomb inside their bellies. This gas is stored safely in their intestines, within storage pouches.

So, when you see manatees swim next time, remember that those teddy bears of the sea rely on farts to swim.

China is breeding 1,100 pound pigs that are as heavy as polar bears.

Go big or go home. It seems that someone in China has taken this a little too seriously.

Deep in the most southern regions of China, there are stories of pigs as big as bears. And, it's no joke. They are actually that big - 1,100 pounds to be exact. It was estimated that there would be a pork shortage in China, so one man decided to take matters into his own hands and plump up the production. Since it's difficult to breed more pigs, it seemed easier to just make the ones they already have bigger. One farmer stated they want to make them as big as possible. I guess that in China, bigger really is better.

Pigs are a smart, but fairly gross animal. And an 1100 pound pig? Well, that's even grosser. It's cool that you can actually ride around on a pig that size, but that's an awful lot of pig poop.

Some snail shells are...hairy?

The slug life is never easy. For example, a crab can leave its shell at any point and just crawl into a new one.

Easy-peasy. There are actually designer shells to be bought for these lucky crabs! However, snails don't have this luxury. They grow their own home, their little piece of armor which keeps them safe.

What's interesting to note is that some snails have hair on their shells. Seems silly, right? What could they possibly use hair for? Well, scientists have tried to find the solution to this Mother Nature puzzle, and they reached one conclusion. It's possible that this hair-like structure of the outer surface is used to help a snail crawl through places which are wet and slippery.

Octopuses taste things with their arms.

Imagine going to a restaurant, and touching the food with your hands, to see what it tastes like. Well, that's exactly what octopuses are doing.

They have what scientists call a touch-taste sense, activated through their sensory receptors which are found on their suckers. When they find prey, they quickly wrap their tentacles around it and pull it in closer.

Being that octopuses are basically one big muscle, this is how they explore the world around them - by tasting it all the time. It's easy for us to smell particles in the air, or taste sweet, bitter, sour, salty and meaty with our tongues. But, it's a bit trickier to do that in the ocean. So octopuses needed to level up their game and they did it perfectly!

Some frogs choose elephant poop piles to be their homes.

Why did the frog cross the grasslands? To jump into a pile of elephant poop, of course.

Elephants are gluttons, or in other words, they eat a lot. Like, really a lot. And, they digest their food very slowly, which means that they poop out most of the stuff they've eaten almost intact. This is why their poop piles are an excellent source of minerals.

This is where frogs come in. They happily jump into this buffet filled with insects and all kinds of other nutrients. It's a whole miniature ecosystem in itself, that provides the frogs with not only food and comfort, but also shelter from the sun.

So you see, in nature everything is connected. But who would have thought it was THIS connected?

Giraffes have solid black, 20 inch tongues.

So, why are their tongues black? This is one of the most frequently asked questions about giraffes. One of the theories, and the most obvious one, is that they are darker in color, black, blue or sometimes purple, because this prevents the tongue from getting sunburned.

How in the world does a tongue get sunburned? Well, just look at the giraffe's body and all will be clear. A giraffe uses its tongue to grab leaves from trees, which means that its tongue is out in the sun a lot. Also, it is 20 inches long, because a giraffe uses it to maneuver and work around any possible thorns that are found around the leaves. Since giraffes are grazers, their tongues get a lot of sun exposure. The dark color helps reflect the sun and keep the tongue from getting a sunburn.

Amynthas worms will jump and go crazy to avoid danger.

These worms are officially called amynthas worms, but they're also known as jumping worms or crazy worms. That's because when you pick them up, they start going all crazy, wriggling and jumping, and even detaching from their own tail.

Yes, you read that right. These little fellas, which are usually sold for bait and used for compost in our gardens, act more like a threatened snake than a little worm when in danger. And, sometimes, they don't mind leaving a piece of themselves if that means that the rest of their body will get to live.

Vampire bat saliva contains stuff that prevents it's victim's blood from clotting.

Why don't vampire bats like mosquitos? Too much competition. Although, mosquitos are mere amateurs when it comes to sucking blood. They prick you and suck just a little. They're more annoying than anything else. But, vampire bats make it their business to do it right.

Here's how. There's this thing called Draculin (yes, it's named after the famous Count Dracula) which is basically a glycoprotein that can be found in vampire bat saliva. It works as an anticoagulant, so that when a vampire bat bites its prey, their blood continues to flow freely so that the vampire bat can drink it. Like sucking juice through a straw. Neat, right?

Horned lizards shoot blood from their eyeballs.

Horned lizards take the term bloodshot eyes to a whole new level. They actually shoot blood out of their eyes. But, in their defense, this is a very *special occasion* kind of defense, not pulled out just for any old predator. If presented with imminent danger, horned lizards would first turn to camouflage for defense. Then, they would turn on their freeze response. They might also try to run, but that rarely works. So, when all else fails, they do the good old blood in your eye trick.

They do this by actually allowing the predator to take them into its mouth. Because of the horns on the lizard's head, it's assumed the predator won't bite on them too hard initially. Then, the lizard squirts the blood right into the predator's eye. They usually react by drooling and shaking their head and this of course, allows the lizard to rush off to safety.

Ladybugs can be cannibals.

Can you imagine something as small and cute as a ladybug having the appetite of a full grown cow? Amazing, isn't it?

Ladybugs are welcome in every garden, because they eat all sorts of insects, mites and aphids. A fully grown

ladybug can eat as many as fifty aphids a day, which boils down to about 5,000 aphids in a ladybug lifetime. (Aphids are small sap sucking insects.)

However, what happens when there is a shortage of aphids? It seems that these sweet little beetles become less sweet when you find out they practice cannibalism. Basically, when there's no food, they will eat whatever they can find just to survive. A grown ladybug won't refrain from eating newly emerged siblings or children, as long as they are still soft-bodied.

Also, some scientists claim that ladybugs sometimes lay infertile eggs on purpose, so that the young ones have ready food when they hatch. Talk about survival, huh?

When Scorpions shed their tails, they will eventually die from constipation.

Lizards can remove their own tails when in danger. Spiders do the same with their legs. This is called autotomy - having body parts which can be easily torn away from the rest of the body, and then, they are replaced by new ones shortly. Sounds like a great skill, no?

Well, it actually depends on what the trade-offs are. If your tail or leg grows back, then sure. Sounds swell. But, if your butt is located at the end of your tail and you just lost it...you're in pretty big trouble.

This is what happens when a scorpion leaves its tail behind. It is also leaving the final parts of its digestive tract.

Seeing that this tail will not grow back, that means that this particular scorpion will never be able to poop again. Ever. And, that also means that it's living on borrowed time, which usually lasts around 8 months.

Most people's cell phones are dirtier than their toilets.

What's the one thing that goes with people everywhere they go? I'll give you a hint. People take it to the bathroom to check out some photos while they're doing their business. They take it to the kitchen to read some comments or watch some videos while they eat. Yup. It's the cell phone.

And, because it goes everywhere with people, it also touches everything they touch. And, that's a whole lot of things and counters and surfaces. Of course, people wash their hands. But, do they wash their phones? Scientists did a study, and they found fecal matter on 1 out of every 6 smartphones! That is one super dirty phone right there in your pocket if you have one.

So, just make sure to clean that mobile germ device as often as you can, because your health depends on it. Or if you really want a phone and your parents won't give you one, that may be okay. You can tell your friends they probably have poop on their phones.

People produce 26 gallons of sweat every year when they sleep.

We all like to sleep alone. More room in the bed, especially if you're tossing and turning all night long. But, what if I told you that, despite what you may think, you aren't really sleeping alone every night in your bed? Oh no. You've got a whole array of bed buddies, such as bacteria, fungi, pollen, lint, soil, sweat, skin cells, and much more.

Now, why do all these things exist in your bed? The answer is simple. Humans produce about 26 gallons of sweat every single year during sleep. And, we all know that moisture is a great enhancer of all sorts of bacteria and fungi. Not to mention dust mites.

So, make sure to wash your sheets on a weekly basis, and maybe, you'll have less bed buddies.

There are 1.5 million dust mites with you in bed each night.

This may make it hard to fall asleep tonight, knowing that you're not alone in your bed sheets. You've got about

1.5 million little friends in there with you. And those are just the dust mites. These are tiny little bugs that live in our beds and bed linens. Their favorite meal is our dead skin cells, so there's always lots to feast on, and because there's always lots to feast on, dust mites multiply at an alarming rate.

How do you fight these nasty little critters? Despite what your parents claim, don't make your bed immediately when you get up. Expose them to sunlight and fresh air, instead of trapping them under the sheets and blankets, and this will destroy most of them.

Is human pee in cigarettes?

Many people want to quit smoking, but claim they can't. This is a great reason to help them do so. Urea is found in some brands of cigarettes. It's put there on purpose to make the flavor more intense. You could do the same by peeing on your apple tree to make the apples taste sweeter. This is because urea is a chemical that is in human urine.

But it's one thing to pee on a tree and another to inhale something that comes out of your own body as a gross byproduct, something you don't need anymore.

So, the next time someone you know says they want to quit smoking but they can't, just tell them about this. If they know they're smoking human pee, maybe they'll be able to drop such a gross habit!

Your feet produce 32 tablespoons of sweat...every day. Change those socks!

We all know that feet smell. We've all been convinced of this numerous times. But, did you also know that feet sweat?

Yes, they certainly do. They do a whole lot of work, after all. It's only natural. Now, there are around 500,000 sweat glands on your feet alone, and these work extra hard sometimes, producing as much as 32 tablespoons of sweat every day. Yes, you read that right!

The stinkiness is caused by bacteria while wearing shoes and socks. Your socks do a great job of soaking up all that sweat during the day, but this is also partly why your feet are so smelly.

Everybody farts around 14 times per day, even your teachers and parents!

Even though we're reluctant to let one rip in public, everyone farts. And, I do mean everyone. The boys, the girls, the men, the women, the old, the young, kings and queens. Every single person in the world farts. And, we do it on average 14 times a day, although it mostly depends on what you eat. So you could go down to 10 times or go as high up as 25. That's a lot of ripping!

Farts are little bubbles of air trapped in our bodies, a mixture of air, carbon and gasses in your intestines. The smelly part is actually only 1% of the whole gas bubble, called hydrogen sulfide.

And, while you may be tempted to hold in a fart, don't. Scientists say it's unhealthy and it may give you a bloated feeling in your stomach.

Doctors investigated a man who was having stomach pains. It turned out that he had eaten 116 nails.

We've all eaten weird foods at some point. Don't be ashamed to admit it. We all have that one thing we like that our friends think is...odd. However, we'll bet your favorite snack isn't as strange as this.

A 43-year old man from India was in the habit of eating nails. Dirty, rusty nails. No one in his family knew it. So, imagine their surprise when he ended up at the hospital and they took out approximately 116 nails, each about 2.5 inches long! Luckily, the good doctors removed the nails from the man's intestines and he was fine eventually. If he wants iron in his snacks, he should just eat more spinach and broccoli.

Everyone has a unique tongue print, just like fingerprints.

You probably already know that fingerprints are unique, being used to catch bad guys both on TV and in real life.

But, did you also know that tongue prints are also unique and no two people have the same tongue print?

Our tongues do a great job at pushing food around in our mouths or pronouncing even the trickiest letters and words. But they also have special ridges which make out their one of a kind print.

Despite this, it's doubtful that tongue prints would ever be used in a crime detection kit, mostly because it's a pretty pricey procedure to collect them and store them properly.

More germs are spread by shaking hands than kissing.

Pucker up! We are surrounded by germs everywhere we go. That's just life and all very normal. Nothing to worry about.

But, if you don't wash your hands properly, you may spread more germs than you're aware of. Researchers have conducted studies, and they realized that handshakes spread more germs (like, the flu and other stomach bugs) than kisses.

How? Because shaking hands is a physical contact (kissing is, too, but you don't kiss everyone you meet, do you?) and you can never be sure what the other person's been touching before they touched you. We touch more things with our hands than we do with our lips. This is how germs are transferred from and between people and surfaces. So, make sure to wash those hands properly!

During your lifetime, you will produce enough spit to fill up two swimming pools.

Our bodies are so weird and wonderful at the same time, but everything they do has a very specific purpose. For example, we have saliva (also known as spit) to help us with food digestion and to keep our mouths clean, but that's not all. If you cut yourself a little, and you've got nothing else lying around, you can spit on it. (I know, gross, but bear with me here.)

Saliva has antiseptic qualities, consisting of great stuff like water, antibacterial compounds (kind of like medicine), mucus, enzymes, and so on. And, you've got more than enough of this spit to go around. During an average human life, a person will produce enough saliva to fill up two big swimming pools!

You have over 59,000 *MILES* of blood vessels in your body.

The human body is so weird and one of the most amazing things in nature. It fits a crazy amount of stuff inside, all with a specific purpose. That is how there are about 59,000 miles of blood vessels in the bodies of kids, whereas the body of an adult houses a whopping 100,000 miles!

Just to put it in perspective, going all the way around the entire earth is about 25,000 miles. This means we could wrap our blood vessels around the earth as many as four times!

This is because these blood vessels are very small, measuring about 5 micrometers. A single strand of human hair is usually around 17 micrometers. So a blood vessel is SUPER tiny.

If a human head is cut off, it remains conscious for 15-20 seconds.

Human history is actually filled with cut off heads. The French were particularly fond of the guillotine, and that is exactly where this idea comes from. The guillotine was a device that when someone was to be killed, they would put their head in and a big blade would come crashing down and quickly cut off their head.

There is a theory called The Astonishing Hypothesis by a molecular biologist named Francis Crick, who stated that everything we think of, our music preferences, beliefs in fairies, or even the way we perceive the color red, comes from electrical activity inside our brains.

At the very basis of our thoughts and everything we think about are chemicals known as neurotransmitters.

They create electrical signals, which is how neurons in our brain communicate, and eventually we get to experience both physical and mental sensations which make up our entire lives.

And strangely enough, once a human head is detached from the body, some electrical activity is still generated inside the brain for anywhere between 4 to 20 seconds. But, that doesn't mean the head knows what happened to the body. Imagine that!

Stomach acid is strong enough to dissolve metal. *(But don't eat it!)*

In order to digest food, our stomachs create acid which dissolves things and reduces food to small tiny pieces. This acid is really strong and does the job perfectly. It can break down any kind of food. And, even more so, it is strong enough to dissolve different metals! But, don't try this at home. Metal will really hurt going down to your stomach and put you in the hospital. It's a REALLY bad idea to eat anything that isn't food.

By the way, how do our stomachs handle this powerful acid? It's actually handled really easily. A new stomach wall is created every three or four days, to protect the rest of the organs. That body of ours... it sure is amazing. But plastic toys will pass right on through in your poop as long as they're small enough. Plastic won't dissolve like metal can. But dissolving metal can burn a hole in your stomach lining. Ouch! Let's just keep food in there, okay?

The human heart is so strong, it can squirt blood 30 feet into the air.

What is the primary muscle that keeps you alive? The heart, of course. It works non-stop to keep you alive. In fact, it beats approximately 100,000 times a day. That's a whole lot of ticking!

Because it pumps so much blood (imagine a faucet on full blast throughout the course of 90 years!), the contractions it creates are under a heavy dose of pressure. In fact, your heart can squirt blood 30 feet into the air! That is only if a hole opened up in your body. But that doesn't happen, I promise.

It's just interesting how powerfully your heart is beating and what it's capable of. Be sure to thank it for all it's work every now and then.

1 in every 2000 babies is born with a tooth already in its mouth.

We all love babies. We love those super cute, toothless grins. But, every once in a while (around 1 in every 2000 or so babies) a baby is born with teeth. Yes, real teeth inside their little mouths. These are called natal teeth, and they're different from neonatal teeth, which come later on.

This doesn't happen that often, but when it does, it's not a cause for alarm. The mom just has a little nibbler on her hands, sooner than she expected to.

Humans can make up to 7,000 different facial expressions.

Humans mostly talk or use verbal communication to convey messages. It's the easiest way, right? Just say it!

But, it's actually impossible not to include our whole body in the communication. So most of the time, non-verbal communication can tell us as much or more than words can.

Body language can tell us if someone is lying, or bored or just concentrating. How? With our facial expressions.

There are approximately 7,000 of them and the stuff they show or communicate is amazing. You crumple your nose, your lips tremble, your eyebrows are squeezed together, your eyelids get tight, your nostrils flare... the list is endless. So, turn that frown upside down, sideways, and every other way.

Some of the atoms in our bodies are stardust that is billions of years old.

You may know that old saying that we are all made of stars. Well, it turns out it's really true. Subatomic particles create atoms. Hydrogen was the first to form. Then, helium, lithium, beryllium, and so on. You get the idea. These are the building blocks that all together make up our planet Earth.

Scientists say that about 40,000 tons of these particles (or stardust from exploded stars) fall down on Earth yearly. It's so tiny we can't see it, but it's there: in the dirt, the plants, animals and of course, in us.

You have all of these inside of you. Blood needs iron. Muscles need sodium and potassium. Your body needs hydrogen, oxygen, carbon, etc. This means that you wouldn't be able to function without stardust. Pretty neat, huh?

And...you might actually have older atoms inside you than your grandpa does!

A sneeze travels 100 miles an hour, faster than a cheetah.

I don't know about you, but I sure don't like it when someone next to me sneezes without covering their mouth.

Why? Because every sneeze is like a miniature germ bomb.

A sneeze is basically how your body gets rid of harmful bacteria and irritants such as allergens and dust. The body ejects them out of your mouth and nose. While sneezing seems like it comes effortlessly, it's not that easy to do. It's like the force of a slingshot. Your body (with the help of your brain) uses as much strength as possible to eject those germs out. The momentum is a staggering 100 miles per hour!

Each sneeze releases thousands of droplets out of your mouth and nose out into the air. So, make sure to always cover up your mouth and nose when sneezing, otherwise you leave miniature boogers all over everything!

You will shed 105 pounds of skin in your lifetime.

Do you know what your body's largest organ is? It's your skin. It's made up of a lot of helpful things, like water, minerals, proteins, lipids, etc. It's main purpose is to protect your body.

In order to do this properly, your skin sheds. Think of a snake shedding its skin, only you don't take it off like a jacket all at once like a snake does.

Human skin sheds particles. About 600,000 particles every hour, to be exact. And, we don't even notice it! This stacks up to about 1.5 pounds a year, and an amazing 105 pounds over a whole lifetime. That's a really big pile of skin.

The strongest muscle you have is your tongue.

Tongues do a lot for us. They twist. They slurp. They roll. They bend. They curl. They taste. You get the idea.

And, while you can't really expect to pull a truck with your tongue, they are still pretty amazing and powerful.

Tongue experts (yes, this is a thing) say that the tongue is the strongest muscle because of two things. First of all, it's not one muscle but a combo of eight separate muscles. Second, its strength is shown through its stamina.

How? Well, let's think about it. When was the last time you felt your tongue was too tired to move food in your mouth or pronounce words? Never, right? It doesn't get tired. Ever.

Skin makes up 15% of your body weight.

If you could take a wild guess at the weight of your skin, how much would it be? Five percent of your overall weight? Fifty? The real answer is 15 percent. Your skin is the heaviest organ you have. Surprising, right?

The skin is made of fleshy stuff. You have the epidermis, the dermis and the subcutaneous tissue. The thickness of these depends on your overall body fat, so of course, skinnier people will have a lighter weighing skin. Your skin, hair, finger and toenails are all part of the integumentary system that helps to protect the body.

Dead people can get goosebumps.

This one pretty much speaks for itself. And, once again, it is not for the faint of heart. Yes. Dead bodies get goosebumps. But, first, let's see how we get goosebumps. It usually happens when we're cold or afraid. The hair simply stands on edge. Do corpses get cold or afraid? Of course not. So, what gives?

When someone dies, something called rigor mortis kicks in. The muscles contract, which means that the body becomes stiff. The muscles underneath the hair follicles also stiffen. Then, when these exact same muscles contract again (as a result of the process of decomposition), the hair pops up, giving the impression of goosebumps. Honestly, that gives ME goosebumps!!!

Dead skin makes up a billion tons of dust in the earth's atmosphere.

So, we've already established that your skin sheds particles. It's your largest organ. It protects your internal organs and bones. It's 15% of your overall body weight. So, needless to say, our skin does an amazing job, day in and day out.

It's no wonder then that dead skin amounts to approximately a billion tons of dust in the earth's atmosphere.

Does this mean that you are breathing in everyone else's dead skin? Sort of!

Also, some researchers say that over half the dust in your home is from skin cells. Well, actually a combination of skin cells, pollen, hair, other dust, materials from your local environment, paper fibers, etc. So, always remember that you are an active part of the air around you, whether you like it or not!

There's an Amazon tribe that eats the cremated bones of their dead.

When a member of the Yonomamo tribe in Brazil and Venezuela dies, other tribe members practice the ritual of endocannibalism. The body is wrapped in cloth and allowed for the soft tissue to decompose for 30-45 days.

The bones are then burned into ashes. The ashes are mixed with bananas and made into a soup that the whole village consumes. This ritual is said to strengthen the community and keep the spirit of the dead tribe member alive.

Think about that the next time you want to complain about eating vegetables!

A museum once had a cheese display made with microbes collected from celebrities.

If you're holding onto a sandwich right now, I suggest you put it down. Being the ultimate fan means different things for different people. Sure, it's fine to get one your favorite basketball player's jersey, but imagine someone making cheese out of their armpit or belly button bacteria? I think we all know the answer to this one.

However, there was an exhibit at the Victoria and Albert Museum in London where there were five types of cheese made from microbes that had been gathered from British celebrities. Bacteria was collected from armpits, ears, noses and belly buttons, and then it was developed further in a lab, where it got ready for cheesemaking.

Ready for that sandwich now?

It is illegal in England to handle salmon in suspicious circumstances.

We've all read funny laws that were created a hundred years ago and are irrelevant for our times today. For example, Arizona passed a law in 1920 that made it illegal for a donkey to sleep in a bathtub. So a really tired donkey that's looking at an old bathtub in the field better think twice!

Not all that long ago, there was a five hour debate at the House of Lords in England, and the decision was unanimous. Adequately named the Salmon Act of 1986, this law outlines what is legal for salmon fishing and also prevents "handling salmon in suspicious circumstances."

While this may sound like it's to prevent the sale of bad or spoiled fish, or that the salmon is up to sneaky behavior, it is actually just to regulate illegal fishing. But it does make the whole thing sound a bit fishy!

In North Korea instead of flower girls and ring bearers they have a hen with a flower in its beak and a rooster with a hot chile pepper walk down the wedding aisle.

Even though many countries have adopted new and modern ways and traditions, some countries still choose to remain faithful to their tradition. North Korea is one of them.

During a traditional North Korean wedding, a rooster and a hen are gently nestled in blue and red cloths, and then placed on the ceremonial table. Then, guests proceed to put flowers and dates into the hen's beak. This symbolizes a beautiful, blessed life for the newlyweds. The rooster gets something much spicier, the poor guy. He gets stuffed with red chilli peppers.

There are over 200 dead bodies laying around on Mt. Everest who did not make it up (or back down).

Mt. Everest has always been the landmark that symbolized people's desire to conquer nature. While it all sounds great to climb and have awesome bragging rights, you should know that an incredible amount of people have died climbing it.

People have fallen into deep cracks in the ice, they've suffocated due to lack of oxygen or they were smashed by falling boulders. Yikes.

But, that's not all. Not only have many people died, but their bodies are still there! Apparently, there are so many of them, more than 200 to be exact, that they sometimes serve as trail markers!

Some common food specialties around the world include pigs feet, horse meat, and cow brain.

Traditions and culture vary throughout the world and this can often be seen with the things we eat. Common dishes in some parts of the world might make us gag, but it's all in what you are used to being served.

For example there's the French *boudin noir*, otherwise known as blood sausage, which is purplish sausage made of pork and pig's blood. In some places in Europe and Asia it would not be unusual to find horse meat on the menu. In France it is often served as steak tartare, ground horse meat mixed with onions, capers, mustard and Worcestershire sauce.

Pigs feet, or sometimes called pig's trotters are often slow cooked to make the meat tender and can be used to make stock and gravy. Cow brains or *cervelle de veau* is a delicacy in Europe and Morocco. It is often served with a side of beef tongue or mixed with scrambled eggs. Yum!

A special food dish in Peru is guinea pigs.

Guinea pigs are often kept as pets and companions but they were originally domesticated to be used as meat! In the Andes Mountains of South America people have been eating guinea pigs since 2000 B.C.

Known as cuy, guinea pig meat is a delicacy that is high in protein and low in fat. It tastes a bit like other small wild game animals and is typically prepared fried, roasted or grilled, served with potatoes, corn or rice.

Guinea Pig is still a popular dish today in these areas and gaining popularity in other places in the world as well. This is because it is easy and inexpensive to raise. Guinea pigs reproduce at a high rate and require very little space to be kept properly.

In Brazil, people cover their pizza with ketchup.

Adventurous eaters out there may want to give this one a try. Many people in Rio de Janeiro eat condiments, such as ketchup, mustard, mayonnaise, or even a bit of olive oil spread on the top of their pizza. Pizza chefs in the area tend to use sauce very sparingly, so the condiments add both flavor and keep the pizza from being too dry.

And in Belgium, they don't eat ketchup with french fries but go with mayonnaise instead. Different ways that people eat common foods around the world can be really interesting.

Made in the USA
Middletown, DE
24 November 2021

53395254R00070